Aggregating clones, colors, equations, iterates, numbers, and tiles

Edited by
János Aczél

Birkhäuser Verlag
Basel · Boston · Berlin

Reprint from *aequationes mathematicae*
Volume 50 (1995), No. 1/2

Editor:

János Aczél
Department of Pure Mathematics
University of Waterloo
Waterloo, Ontario
Canada N2L 3G1

A CIP catalogue record for this book is available from the Library of Congress, Washington D.C., USA

Die Deutsche Bibliothek – CIP-Einheitsaufnahme

Aggregating clones, colors, equations, iterates, numbers, and tiles/ed. by János Aczél. – Basel ; Boston ; Berlin:
Birkhäuser, 1995
  ISBN-13:978-3-7643-5243-1   e-ISBN-13:978-3-0348-9096-0
  DOI:10.1007/978-3-0348-9096-0

NE: Aczél, János [Hrsg.]

© 1995 Birkhäuser Verlag Basel, P.O. Box 133, CH-4010 Basel
Printed on acid-free paper produced of chlorine-free pulp ∞

ISBN-13:978-3-7643-5243-1

9 8 7 6 5 4 3 2 1

Contents

Preface

The journal *aequationes mathematicae* publishes papers in pure and applied mathematics and, in particular, articles on functional equations, combinatorics and dynamical systems. Its 50th volume appears in 1995. To mark this occasion, we are publishing in book form a representative collection of outstanding survey papers assembled for our anniversary issue of *aequationes mathematicae*.

The articles by Quackenbush, Targonski and Moszner discuss composition of functions from different points of view: universal algebra, dynamical systems (iteration) and functional equations. The Ono-Robbins-Wahl and the Vince papers, on number theory and tiles, respectively, are thematically linked by lattices. Combinatorics, in turn, links the Vince paper with that of Tutte, whose subject is chromatic sums, its tools differential and functional equations. The Paganoni-Rätz and the Forti papers deal with conditional functional equations and with the related topic of stability. Applications to the social and behavioral sciences, in particular to aggregation (and some theory) are presented in the paper by J. Aczél.

The aim of the collection is to survey selected fields of current interest. We trust that it will be useful and informative for researchers, teachers, graduate and advanced undergraduate students of mathematics, and for those interested in applications in related fields.

János Aczél

Aequationes Mathematicae **50** (1995) 1
University of Waterloo

0001–9054/95/020001–01 $1.50 + 0.20/0

Editorial

Volume 50 of Aequationes Mathematicae

This is the fiftieth volume of *aequationes mathematicae*.

Not only our modesty but also lack of space keeps us from self-congratulation. We still celebrate the occasion with a modest collection of carefully selected survey papers and with a Cumulative Index of the first 50 volumes of our journal.

The Editors

Aequationes Mathematicae **50** (1995) 3–16
University of Waterloo

0001–9054/95/020003–14 $1.50 + 0.20/0
© 1995 Birkhäuser Verlag, Basel

Survey Papers

A survey of minimal clones

Robert W. Quackenbush

Dedicated to the memory of Trevor Evans

Summary. A clone is a set of composition closed functions on some set. A non-trivial fact is that on a finite set every clone contains a minimal clone. This naturally leads to the problem of classifying all minimal clones on a finite set. In this paper I survey what is known about this classification. Rather than repeat the arguments used in the original papers, I have tried to use known results about finite algebras to give a more coherent and unified description of the known minimal clones.

1. Introduction

A function f is of *finite arity* on a finite set S if for some natural number $n, f\colon S^n \to S$; here n is the *arity* of the function f (by convention, a 0-ary function is a constant). In particular, there are the *projection* functions $p_i^n\colon S^n \to S$ defined by $p_i^n(a_1, \ldots, a_n) = a_i$ for all $(a_1, \ldots, a_n) \in S^n$. The *composition* of functions is defined by: if $f\colon S^n \to S$ and $g_i\colon S^m \to S$ for all $1 \le i \le n$, then $f(g_1, \ldots, g_n)\colon S^m \to S$ where for all $(a_1, \ldots, a_m) \in S^m$,

$$f(g_1, \ldots, g_n)(a_1, \ldots, a_m) = f(g_1(a_1, \ldots, a_m), \ldots, g_n(a_1, \ldots, a_m)).$$

A *clone* on a finite set S is a set, $\mathscr{C}$, of functions of finite arity on S containing the projection functions and closed under composition.

A clone on S is *minimal* if it is an atom in the lattice of all clones on S (the set of all clones on S is closed under arbitrary intersections, and ordering is by containment). Obviously, a minimal clone is the smallest clone containing any one of its non-projections (i.e., a function which is not a projection). We denote by $[f]$

AMS (1991) subject classification: Primary 08A02, Secondary 08B05.

Manuscript received November 28, 1991 and, in final form, March 29, 1993.

the smallest clone containing the function f; all minimal clones are of the form $[f]$ for some function f. More specifically, the clone $[f]$ is minimal iff for any non-projection $g \in [f]$, we have $f \in [g]$.

EXAMPLE 1.1. The 2-element semilattice $\langle \{0, 1\}; \vee \rangle$ (where $a \vee b$ is the larger in the natural order) is minimal (i.e., $[\vee]$ is a minimal clone on $\{0, 1\}$). This is because given any semilattice term in the variables $x_1, \ldots, x_n$, if we set $x_i = x_2$ for $i \geq 2$, then the term reduces to $x_1 \vee x_2$.

In fact, this argument actually shows that every non-trivial (i.e., having more than one element) semilattice generates a minimal clone; this includes infinite semilattices too, but in this survey we are only interested in minimal clones on a finite set. This is a general phenomenon: if the algebra $\langle A; f \rangle$ generates a minimal clone (i.e., $[f]$ is a minimal clone on A) and $\langle B; g \rangle$ satisfies the same identities as $\langle A; f \rangle$ then $\langle B; g \rangle$ also generates a minimal clone. Let us say that a clone $\mathscr{C}$ has *arity* n if it contains a non-projection, f, of arity n but contains no non-projection of lesser arity; in particular, this means that f is *essential*, i.e., it depends on all its variables. Notice that, if a clone has arity at least 2, then every function t in the clone is *idempotent*: $t(x, \ldots, x) \approx x$. Thus, the search for minimal clones of arity n reduces to the search for varieties, $\mathfrak{A}$, with one n-ary operation f such that for any non-projection $t \in [f]$, we have $f \in [t]$. Notice that for each such t this is equivalent to a certain identity in n variables being true in $\mathfrak{A}$. Does this mean that $\mathfrak{A}$ can be defined by identities in at most n variables? For the case of semilattices, call a groupoid a 2-*semilattice* if it satisfies all semilattice identities in at most 2 variables; then any non-trivial 2-semilattice generates a minimal clone. The variety of 2-semilattices is very large and quite interesting; see for instance [4] for an analysis of a small part of it.

Unfortunately, things are not always this nice. It may be that for some $m > n$ there is an m-ary term such that $t(x_1, \ldots, x_m) \approx x_1$ is true in $\mathfrak{A}$; such an identity is called an *absorption identity*. But this is an identity in more than n variables and so need not be a consequence of the n-variable identities of $\mathfrak{A}$.

EXAMPLE 1.2. A *rectangular band* is an idempotent semigroup satisfying $xyz \approx xz$. Every rectangular band is of the form $\langle S \times T; \cdot \rangle$ where $S \times T$ is the direct product of (non-empty) sets S and T, and where $(a, b) \cdot (c, d) = (a, d)$. Call a rectangular band *left-trivial* (*right-trivial*) if it satisfies $xy \approx x$ ($xy \approx y$); in these cases the semigroup operation is a projection and so the clone generated by any left-trivial (or right-trivial) rectangular band is the trivial clone of all projections. If a rectangular band is neither left- nor right-trivial then it generates a minimal clone since all terms which are not projections are of the form $x_i \cdot x_j$ for $i \neq j$, from which the original operation is obviously recoverable. Can we, in analogy with 2-semilattices, conclude that any groupoid which satisfies all rectangular band identities in at most 2 variables and which is neither left- nor right-trivial generates a minimal clone? The following counterexample of B. Csákány [3], shows that the answer is no. Notice that in rectangular bands the term $t(x, y, z) = (xy)(zx)$ satisfies $t(s, y, z) \approx x$. Suppose that we can construct a 2-rectangular band $\langle S; \cdot \rangle$ in which $t(x, y, z)$ is not a projection; then $[t]$ will contain no binary functions which are not projections (since any 2-generated subalgebra will satisfy $t(x, y, z) \approx x$) and so will not contain $x \cdot y$: i.e., our 2-rectangular band does not generate a minimal clone. Let $|S| \geq 3$; define $\cdot$ on each 2-element subset of S to be either left- or right-trivial with at least one 2-subset being left-trivial and one being right-trivial. This yields a 2-rectangular band (since every 2-generated subalgebra is either left- or right-trivial) which is neither left- nor right-trivial. Let T be an at least 2-element subset of S maximal with respect to all its 2-subsets being right-trivial, and choose $c \in S - T$. Then there is some $a \in T$ such that $\{a, c\}$ is left-trivial; finally, choose $b \in T - a$. But then $t(a, b, c) = (ab)(ca) = bc \in \{b, c\}$, and so $t(x, y, z) \approx x$ fails in S.

This article is a survey of what is currently known about minimal clones, and is aimed at the readers of this journal; however, I will assume that the reader is comfortable with the rudiments of composition of functions (clone theory) and manipulation of identities (equational logic). The lecture notes of Á. Szendrei, [13] are highly recommended, even to those familiar with the rudiments of clone theory. In particular, it has an extensive bibliography. I will emphasize the use of known results from universal algebra which can be used in the search for minimal clones, but will try to keep the paper reasonably self-contained. The reader should be aware of the fact that the approach used here was rarely the original approach; indeed, some of the original research is all but unreadable. The usual approach is to start with some function g and sift through $[g]$ for some function f with some nice properties, and then use these nice properties to prove that $[f]$ is minimal. The sifting process can be quite tedious and will be omitted. But the proofs that a given clone is minimal (distinguishing real gold from fool's gold, if you will) are often quite interesting.

As an example of using known universal algebraic facts to find minimal clones, let us prove Á. Szendrei's characterization of those Mal'cev functions generating a minimal clone [14]. Recall that $d(x, y, z)$ is Mal'cev if $d(x, y, y) \approx d(y, y, x) \approx x$ (the canonical example is $xy^{-1}z$ in groups). For a prime p, $A_p = \langle \{0, \ldots, p - 1\};$ $x - y + z (\mathrm{mod}\, p) \rangle$ generates a minimal clone and $x - y + z$ is obviously Mal'cev. The minimality of A_p is easily proved once you realize that it is the *idempotent reduct* (i.e., the subclone of all the idempotent terms) of the cyclic group of order p. Note that for $p = 2$ the clone has arity 3 while for $p > 2$ the arity is 2. We can now state Szendrei's result.

THEOREM 1.3 (Á. Szendrei [14]). *Let S be a finite set with at least 2 elements and let d be Mal'cev on S. Then [d] is a minimal clone if and only if for some prime p there*

exists an elementary Abelian p-group $\langle S; + \rangle$ with $d(x, y, z) = x - y + z$ (i.e., $\langle S; d \rangle$
belongs to the variety generated by $A_p = \langle \{0, \ldots, p - 1\}; d \rangle$).

Proof. That $\langle S; d \rangle$ generates a minimal clone is easy since, for $p > 2$, d can be recovered from any essentially binary term, while each term is a linear combination of distinct variables whose coefficients sum to 1; consequently, d can be recovered from every term which depends on at least 2 variables. Let $\langle T; d \rangle$ be a minimal non-trivial (i.e., having more than one element) subalgebra of $\langle S; d \rangle$; we first show that $\langle T; d \rangle \cong A_p$, for some p. As d is idempotent, every congruence class of $\langle T; d \rangle$ is a subalgebra. Hence, $\langle T; d \rangle$ is simple and has no proper, non-trivial subalgebras; i.e., $\langle T; d \rangle$ is *plain*. As d is Mal'cev, a theorem of R. McKenzie [6] asserts that $\langle T; d \rangle$ is either *quasiprimal* (i.e., the *discriminator function* $t(x, y, z)$ defined by $t(a, a, b) = b$ and otherwise $t(a, b, c) = a$, belongs to $[d]$) or *affine* (i.e., every $t \in [d]$ satisfies the *term condition*: for all $a, b, a_1, \ldots, a_n, b_1, \ldots, b_n \in T$, $t(a, a_1, \ldots, a_n) = t(a, b_1, \ldots, b_n)$ implies $t(b, a_1, \ldots, a_n) = t(b, b_1, \ldots, b_n)$). Notice that each A_p is affine. If $\langle T; d \rangle$ were quasiprimal then the clone it generates would not be minimal: the clone contains the discriminator function and so contains the *dual discriminator* function $t'(x, y, z)$ defined by $t'(a, a, b) = a$ and otherwise $t'(a, b, c) = c$. To see this, note that $t'(x, y, z) \approx t(x, t(x, y, z), z)$. But the dual discriminator function does not generate a clone containing a Mal'cev function (examine the restriction of t' to any 2-element set), while the discriminator function is a Mal'cev function. Thus, $\langle T; d \rangle$ is affine, meaning that $d(x, y, z) = x - y + z$ for some abelian group on T (the term condition implies that every Mal'cev function is of this form). As $\langle T; d \rangle$ is simple, so is the abelian group on T; i.e., $\langle T; d \rangle$ is isomorphic to A_p for some prime p. To complete the proof, it suffices to prove that $\langle S; d \rangle$ satisfies the same identities as $\langle T; d \rangle$. First note that any absorption identity true in $\langle T; d \rangle$ is also true in $\langle S; d \rangle$ since otherwise some term is a non-projection on S while being a projection on T, and so we cannot recover d from this term, violating the minimality of the clone. Thus, if we can prove that $\langle T; d \rangle$ has a basis of absorption identities relative to $d(x, y, y) \approx d(y, y, x) \approx x$ then we are done. But now invoke the following observation of R. Padmanabhan and B. Wolk [7]: in the presence of $d(x, y, y) \approx d(y, y, x) \approx x$, the identity $s \approx t$ is equivalent to the identity $d(s, t, u) \approx u$, where u is a variable not occurring in s or t. □

2. The existence of minimal clones

QUESTION 2.1. Does every clone on a finite set contain a minimal clone?

One would hope so. But why could we not have a clone such that for every $n \geq 1$ the subclone generated by any essential function of arity $\geq n$ contains no functions of arity less than n which are not projections and contains an essentially m-ary function for every $m \geq n$? Such a clone could not have a minimal subclone. This is ruled out by the following fundamental lemma of S. Świerczkowski [12]. Notice that, for $n \geq 2$, if an essentially n-ary function f generates a clone of arity n, then whenever we equate two variables in f, the function f reduces to a projection; if we always reduce to the same projection, say x_i, no matter which two variables are equated, we will call f a *semiprojection* onto x_i (equivalently, $f(x_{\varrho(1)}, \ldots, x_{\varrho(n)}) = x_{\varrho(i)}$ whenever ϱ is not $1 - 1$).

LEMMA 2.2 (Świerczkowski's Lemma [12]). *Let f be essentially n-ary for $n \geq 4$; if $[f]$ has arity n then f is a semiprojection.*

Proof. We prove it only for $n = 4$; the case $n > 4$ is similar. Case 1. For some i, if we equate all but the ith variable, then f reduces to x_i. We may assume $i = 1$ so that $f(x, y, y, y) \approx x$. But then equating any two of x_2, x_3, x_4 yields x_1. Hence, $f(x, w, y, y) \approx x$ so that $f(x, x, y, y) \approx x$. Now set $x_1 = x_2$. Then we must have $f(x, x, y, z) \approx x$; similarly, $f(x, y, x, z) \approx f(x, y, z, x) \approx x$. Thus f is a semiprojection. Case 2: $f(x, y, y, y) \approx f(y, x, y, y) \approx f(y, y, x, y) \approx f(y, y, y, x) \approx y$. But then for any $i \neq j$, setting $x_i = x_j$ forces f to reduce to $x_i = x_j$. This leaves no variable for $f(x, x, y, y)$ to equate to; Case 2 cannot occur. $\square$

THEOREM 2.3. *On a finite set, every non-trivial clone contains a minimal clone.*

Proof. Let $|S| = n$ and let $\mathscr{C}$ be a non-trivial clone on S. For $k \geq 0$ let us look at the set of all non-trivial k-ary subclones of $\mathscr{C}$ which are generated by a k-ary function; trivially, this set is non-empty. Świerczkowski's Lemma implies that $k \leq n$. But on a finite set there are only finitely many functions of arity at most n. Hence, this set of subclones is finite, non-empty and ordered by containment; its non-empty subset of minimal members is the set of minimal subclones of $\mathscr{C}$. $\square$

3. Rosenberg's type theorem

The only minimal clones of arity 0 are those in the variety of *pointed sets* (i.e., sets with a unique distinguished element). Unary minimal clones satisfy either $f^2(x) \approx f(x)$ or $f^p(x) \approx x$ for some prime p; see R. Pöschel and L. A. Kalužnin [10] (but this is easy to prove in any case). Binary minimal clones are given by idempotent groupoids, and as we have seen, are quite varied.

QUESTION 3.1. What is the structure of binary minimal clones?

For $n \geq 4$, Świerczkowski's Lemma implies that minimal clones of arity n are generated by semiprojections. For $n = 3$, besides the semiprojections we also have majority functions and the operation $x + y + z$ (mod 2) on $\{0, 1\}$ (and on all its finite powers). Rosenberg's Type Theorem asserts that these are the only possibilities. Recall that a ternary function m is called *majority* if $m(x, x, y) \approx m(x, y, x) \approx m(y, x, x) \approx x$; it is called *Pixley* if $m(x, x, y) \approx m(y, x, x) \approx m(y, x, y) \approx y$, and it is called *minority* if $m(x, y, y) \approx m(y, x, y) \approx m(y, y, x) \approx x$. Let $m(x, y, z)$ generate a clone of arity 3. Then there are 8 possibilities for m upon equating two of the variables in $m(x, y, z)$ in the 3 possible ways; one yields a majority, one a minority, 3 a semiprojection and 3 a Pixley function (modulo a permutation of variables).

LEMMA 3.2 (B. Csákány [2]). *Let f be an n-ary semiprojection (majority function) and let $g \in [f]$ be essentially n-ary (3-ary); then g is a semiprojection (majority function).*

Proof. The proof is by induction on the length of g as a term in f; the flavor of the proof is typical of the subject. Let f be an essential n-ary semiprojection onto x_1. As g is essentially n-ary, its length is at least 1. If the length is 1 then g is just f with its variables permuted and so is a semiprojection. Otherwise, $g(x_1, \ldots, x_n) = f(g_1(x_1, \ldots, x_n), \ldots, g_n(x_1, \ldots, x_n))$. By induction, each g_i is a semiprojection onto $x_{a(i)}$ (either essential or a full projection); hence, g is a semiprojection onto $x_{a(1)}$. $\square$

THEOREM 3.3 (Rosenberg's Type Theorem [11]). *Let $f(x, y, z)$ generate a minimal clone of arity 3 on the set S; then f is either a semiprojection, a majority function or of the form $x + y + z$ where $\langle S; + \rangle$ is an elementary 2-group.*

Proof. If f were Pixley, then $[f]$ would contain the majority function $g(x, y, z) \approx f(x, f(x, y, z), z)$; but by Lemma 3.2, $f \notin [g]$, contradicting the minimality of $[f]$. Thus, we may assume that f is a minority function. By Theorem 1.3, $\langle S; f \rangle$ is in the variety generated by A_p for some prime p. But only for $p = 2$ is $x - y + z$ a minority function. Hence, $p = 2$ and we are done. $\square$

4. Conservative minimal clones

A function $f: S^n \to S$ is *conservative* if for all $s_1, \ldots, s_n \in S$ we have $f(s_1, \ldots, s_n) \in \{s_1, \ldots, s_n\}$; equivalently, every non-empty subset of S is a subalgebra of $\langle S; f \rangle$. Since a composition of conservative functions is conservative, we call a clone *conservative* if all its functions are conservative. In this section we will look at conservative minimal clones. If f is an n-ary conservative non-projection on S then not all algebras in the variety generated by $\langle S; f \rangle$ are conservative: the conservative semilattices are just those whose underlying order is a chain. But this might be to our advantage; could it not be that, if $\langle T; g \rangle$ forms a minimal clone, then the variety it generates is also generated by a conservative algebra? Again this is too much to ask for; for $p > 2$, the variety generated by A_p contains no non-trivial conservative algebras. Perhaps it is possible to characterize when this happens. In any case, conservative functions can be highly complicated, yet their clones are more easily analyzed than arbitrary clones; hence, conservative functions are a good special case to analyze.

All binary and majority conservative minimal clones are known in a very strong manner (there is an efficient algorithm for determining whether a binary or majority conservative clone is minimal), while for $n > 3$, conservative minimal clones are completely characterized but only in an abstract manner. A k-ary conservative function on an n-element set is completely determined by its k-element subalgebras. Moreover, if f is a k-ary non-projection on a k-element subset A of S, then $\langle A; f \rangle$ generates a minimal clone whenever $\langle S; f \rangle$ does. Thus, it suffices to determine the k-ary conservative functions on a k-element set which generate a minimal clone and to determine how to glue these together to form a k-ary conservative minimal clone on a larger set.

For binary functions, the only binary minimal clones on a 2-element set are the two semilattice clones. Thus, if $\langle S; f \rangle$ generates a binary conservative minimal clone then f restricted to each 2-element subset of S is either a semilattice, first or second projection. This means that $\langle S; f \rangle$ satisfies all 2-variable identities common to these three 2-element algebras.

LEMMA 4.1 (B. Csákány [3]). *If there exist a non-trivial $g \in [f]$ and non-isomorphic subalgebras $A, B \in \langle S; f \rangle$, such that $\langle A; g \rangle \cong \langle B; g \rangle$, then $[f]$ is not a minimal clone.*

THEOREM 4.2 (B. Csákány [3]). *If $\langle S; f \rangle$ generates a minimal clone and f is a binary conservative function then some 2-element subalgebra is a semilattice and both projections do not appear as subalgebras.*

Proof. Choose $T \subseteq S$ maximal with the property that every 2-element subalgebra of $\langle T; f \rangle$ is a first projection; we must have $T \neq S$. If no 2-element subalgebra of $\langle S; f \rangle$ were a semilattice then we could proceed as in Example 1.2. To see that both first and second projections cannot occur, note that $f(x, f(y, x))$ is non-trivial on S (since we have a 2-element subsemilattice) but equals x on 2-element sets where f is a first or second projection. Now invoke Lemma 4.1.   □

LEMMA 4.3. *Let* $\langle S; \cdot \rangle$ *satisfy* $xx \approx x$, *and*

$$x(yx) \approx (xy)x \approx x(xy) \approx (xy)y \approx (xy)(yx) \approx xy$$

but not $xy \approx x$; *then* $[\cdot]$ *generates a minimal clone*.

Proof. For $n \geq 2$ take any term $t(x_1, \ldots, x_n)$ in at least 2 variables whose first variable is x_1; then $t(x, y, \ldots, y) \approx xy$. □

We are now ready to state B. Csákány's characterization of binary conservative minimal clones, clothed somewhat more elegantly than in the original.

THEOREM 4.4 (B. Csákány [3]). *Let f be a binary conservative function on S with* $|S| > 1$. *Then* $[f]$ *is a minimal clone if and only if* $\langle S; f(x, y) \rangle$ *or* $\langle S; f(y, x) \rangle$ *belongs to the variety defined by*:

$$xx \approx x, \qquad x(yx) \approx (xy)x \approx x(xy) \approx (xy)y \approx (xy)(yx) \approx xy,$$

but not to the subvariety defined by: $xy \approx x$.

5. Conservative majority functions

Next, let us consider minimal clones generated by conservative majority functions. By Świerczkowski's Lemma and Lemma 3.2, any minimal subclone of a clone on $\{0, 1, 2\}$ generated by a majority function is itself generated by a majority function. Up to isomorphism there are 12 majority functions on $\{0, 1, 2\}$ which generate 3 distinct minimal clones. The functions will be given by their behaviour on distinct triples (x_1, x_2, x_3) since otherwise, majority rule prevails.

The first clone, $\mathcal{M}_0$, is generated by $m_0(x_1, x_2, x_3) = 0$; it is the only majority function in the clone. The second clone, $\mathcal{M}_1$, is generated by any one of $m_i(x_1, x_2, x_3) = x_i$ for $i \in \{1, 2, 3\}$. The third clone, $\mathcal{M}_2$, is generated by $m_4(x_1, x_2, x_3) = x_{i+1}$ if $x_i = 2$ (subscripts mod 3). Notice that m_4 never takes on the value 2 on distinct triples and that interchanging 0 and 1 is an automorphism. There are 8 majority functions on $\{0, 1, 2\}$ with this property and these are exactly the 8 majority functions in $\mathcal{M}_2$; we denote these m_j for $4 \leq j \leq 11$.

THEOREM 5.1 (B. Csákány [3]). *Up to isomorphism, a majority function on* $\{0, 1, 2\}$ *generates a minimal clone if and only if it is contained in* $\mathcal{M}_i$ *for some* $i \in \{1, 2, 3\}$.

How do we glue together copies of these majority functions on $\{0, 1, 2\}$ to construct conservative majority functions on larger sets which generate minimal clones? Our first worry is that on an at least 4-element set a majority function might generate a proper semiprojection. By the next lemma, this cannot happen. For $n \geq 3$, an n-ary function u is *near-unanimity* if

$$u(x, y, \ldots, y) \approx u(y, x, y, \ldots, y) \approx \cdots \approx u(y, \ldots, y, x) \approx y.$$

THEOREM 5.2 (B. Csákány [3]). *If m is a majority function on S and $g \in [m]$ is a non-projection then g is a near-unamity function.*

This lemma, Świerczkowski's Lemma and Rosenberg's Type Theorem imply that any minimal subclone of a clone generated by a majority function is again generated by a majority function. Thus, let m be a conservative majority function on S generating a minimal clone. For each 3-element subset A of S we know that $m|_A$ is isomorphic to some majority function in $\mathscr{M}_i$ for some $i \in \{1, 2, 3\}$. Our next worry is that for 3-element subsets $A, B \subseteq S$ we may have $\langle A; m|_A \rangle \not\cong \langle B; m|_B \rangle$ but both coming from the same $\mathscr{M}_i$ (of course, this cannot happen with $\mathscr{M}_0$ as it contains a unique majority function). This cannot happen for the following reason: it is possible to find a term t such that $\langle A: t|_A \rangle \cong \langle B; t|_B \rangle$; by Lemma 4.1 this contradicts the minimality of our clone. This is our last worry.

THEOREM 5.3 (B. Csákány [3]). *Let m be a conservative majority function on S. Then $[m]$ is a minimal clone if and only if for some $i \in \{1, 2, 3\}$ there is a majority function $m_k \in \mathscr{M}_i$ such that for every 3-element subset $A \subseteq S$, $\langle A; m|_A \rangle$ is isomorphic to $\langle \{0, 1, 2\}; m_k \rangle$.*

Notice that this provides us with a polynomial time algorithm for determining whether a conservative majority function generates a minimal clone.

6. Conservative semiprojections

Finally, we have the case of conservative minimal clones generated by semiprojections. The case of an n-ary semiprojection on an n-element set has been completely settled. For $n = 3$ this was done by B. Csákány [2] where he found all minimal clones on a 3-element set; the case $n \geq 4$ was done by J. Ježek and R. Quackenbush [5]. Since the description (but not the proof) for $n \geq 4$ when applied to the case $n = 3$ yields Csákány's result, I will restrict n to be at least 4. Thus, let $S = \{a_1, \ldots, a_n\}$ with $n \geq 4$ and let f be an n-ary semiprojection onto x_1 on S (necessarily conservative) generating a minimal clone.

LEMMA 6.1 ([5]). *Without loss of generality, $f(x_1, \ldots, x_n) \in \{x_1, x_n\}$.*

Proof. By renumbering the elements of S, we may assume that $f(a_1, \ldots, a_n) = a_n$. Let $g(x_1, \ldots, x_n) = f(x_1, \ldots, x_{n-1}, f(x_1, \ldots, x_n))$. Then $g \in [f]$ and has the required property; since $[f]$ is minimal, this means that $[g] = [f]$. $\square$

This is the first in a long series of lemmas which the authors prove; each adds to the properties that such an f can be assumed to possess. I will bypass these and proceed to the end result. Lemma 6.1 says that f is defined by a binary relation $R \subseteq S^2$ as follows: $f(a_1, \ldots, a_n) = a_n$ if $|\{a_1, \ldots, a_n\}| = n$ and $(a_1, a_n) \in R$; otherwise $f(a_1, \ldots, a_n) = a_1$. The additional properties possessed by f are described in terms of R.

A binary relation $R \subseteq S^2$ is *bitransitive* if R is a reflexive, transitive relation and if the automorphism group of R is edge-transitive (i.e., for (a, b), $(c, d) \in R$ with $a \neq b$, $c \neq d$, there is an automorphism of R mapping a to c and mapping b to d). It is relatively straightforward to see that bitransitive relations fall into two classes. The first class consists of equivalence relations such that all blocks with more than one element have the same size; this is a very simple class. The second class consists of posets of length 1 (i.e., whose longest chains have 2 elements) whose non-trivial connected components are all isomorphic and all have edge-transitive automorphism groups. This last class is extremely complicated and apparently all but uninvestigated. See [1] for a very deep analysis of the case corresponding to linear spaces (i.e., minimal elements are points, maximal elements are lines and two points determine a unique line).

THEOREM 6.2 ([5]). *Without loss of generality, f is defined by a bitransitive relation.*

Do all such functions generate minimal clones? If so, do distinct relations generate distinct minimal clones? The answer to both questions is yes, and the key to proving this is the extreme symmetry of bitransitive relations.

Let R be a bitransitive relation on S and let f_R be the semiprojection defined by R. Denote by $\mathscr{C}(R)$ the set of all functions $g(x_1, \ldots, x_m)$ on S (for $m \geq 1$) such that there is an $i \in \{1, \ldots, m\}$ with:

(1) $(a_i, g(a_1, \ldots, a_m)) \in R$ for any $a_1, \ldots, a_m \in S$;
(2) if $\{a_1, \ldots, a_m\} \neq S$, then $g(a_1, \ldots, a_m) = a_i$;
(3) g is conservative and preserves all automorphisms of R.

THEOREM 6.3 ([5]). *$\mathscr{C}(R)$ is a clone containing f.*

The proof of Theorem 6.3 is a straightforward induction argument.

THEOREM 6.4 ([5]). *Let f be the semiprojection defined by the bitransitive relation R. Then* [f] *is a minimal clone. Moreover, distinct bitransitive relations determine distinct minimal clones.*

Proof. Let $\mathscr{C}$ be a minimal subclone of $[f]$; thus, $\mathscr{C}$ is also a subclone of $\mathscr{C}(R)$. By Theorem 6.2, $\mathscr{C} = [g]$ for some n-ary semiprojection defined by a bitransitive relation $T \subseteq S^2$. By property (1), $T \subseteq R$ and so, by property (3), $T = R$, and therefore $g = f$. □

Consider now the case where the size of S is greater than the arity of f. The characterization is complete but far from efficient (and perhaps hopelessly inefficient). Let f be an n-ary conservative semiprojection on S with $|S| = m > n$ and such that $[f]$ is a minimal clone. The *support* of f is the set of all n-element subsets, A, of S such that f is not a projection on A; a member of the support of f will be called a *base set*.

THEOREM 6.5 ([5]). *Without loss of generality, the restrictions of f to any two base sets yield isomorphic algebras.*

Thus, all base algebras are isomorphic, and so determined up to isomorphism, by a unique bitransitive relation on $\{1, \ldots, n\}$; we will say that f is *uniformly determined* by a bitransitive relation. However, the phrase "up to isomorphism" papers over much unknown territory; it seems that the way in which the isomorphisms are glued together is extremely intricate. In addition, it is not at all clear which sets of n-subsets of S can be the supports of such functions.

THEOREM 6.6 ([5]). *If f is uniformly determined by a bitransitive equivalence relation then the support of f covers S (i.e., every element of S is contained in some base set).*

EXAMPLE 6.7 ([5]). Let f be the 4-ary semiprojection onto x_1 on $\{0, 1, 2, 3\}$ such that $f(a_1, a_2, a_3, a_4) = a_4$ if $\{a_1, a_4\} = \{0, 3\}$ and $\{a_2, a_3\} = \{1, 2\}$. By Theorem 6.4, we know that $[f]$ is a minimal clone. Then $\langle \{(0, 0), (1, 1), (2, 2), (3, 3), (0, 3)\}; f \rangle$ is a subalgebra of $\langle \{0, 1, 2, 3\}; f \rangle^2$, which generates a minimal conservative clone and has exactly three 4-sets in its support.

In contrast with equivalence relations is the following result about order relations.

THEOREM 6.8 ([5]). *Let f be uniformly determined by a bitransitive order relation R, and have exactly one base set M (which is a proper subset of S). Then $[f]$ is minimal if and only if R has only one non-trivial connected component and that one component is the 2-element chain.*

That $[f]$ is minimal implies that R is as described, follows (as usual) by finding appropriate functions in $[f]$. That such an R makes $[f]$ minimal follows from the following result whose proof is not trivial. An identity of the form $f(x_1, \ldots, x_n) \approx x_i$, for some i, is called an *absorption* identity.

THEOREM 6.9 ([5]). *The algebras $\langle S; f \rangle$ and $\langle M; f \rangle$ satisfy the same absorption identities.*

In fact, we have a rather nice characterization in terms of absorption identities. However, it is very far from the practicality of Theorem 5.3.

THEOREM 6.10 ([5]). *Let f be an n-ary conservative semiprojection on S uniformly determined by the bitransitive relation R on $\{1, \ldots, n\}$. Then $[f]$ is a minimal clone if and only if $\langle S; f \rangle$ satisfies all the absorption identities of $\langle \{1, \ldots, n\}; f_R \rangle$.*

Proof. One direction is clear: if $\langle S; f \rangle$ fails some absorption identity true in the algebra $\langle \{1, \ldots, n\}; f_R \rangle$, then $[f]$ cannot be minimal as the term corresponding to the absorption identity is trivial on $\langle \{1, \ldots, n\}; f_R \rangle$, but not on $\langle S; f \rangle$. Conversely, let $\langle S; f \rangle$ satisfy the same absorption identities as $\langle \{1, \ldots, n\}; f_R \rangle$, and let $g \in [f]$ generate a minimal clone. By Theorem 6.5, we may assume that g is uniformly determined by some bitransitive relation T. If $m > n$, then $g = x_1$ would be an absorption identity true in $\langle \{1, \ldots, n\}; f_R \rangle$ but not in $\langle S; f \rangle$; hence, $m = n$. But as $\langle \{1, \ldots, n\}; f_R \rangle$ generates a minimal clone, we must have $T = R$ and $g_R = f_R$. As each subalgebra on a base set is isomorphic to $\langle \{1, \ldots, n\}; f_R \rangle$, $f - g$ on all base sets. On any n-set not a base set, we have $f = x_1$, and so $g = x_1 = f$. Hence, $f = g$ and $[f]$ is minimal. $\square$

7. The Pálfy clones

In [8], P. P. Pálfy constructs a k-ary minimal clone on an n-element set for every $3 \leq k < n$: define k-ary f on $\{1, 2, \ldots, n\}$ by $f(a_1, \ldots, a_k) = k + 1$ if $a_1 = 1$ and $\{a_1, \ldots, a_k\} = \{1, \ldots, k\}$, and otherwise a_1. The proof that $[f]$ is minimal uses the following lemma; for any term t, let $\sigma(t)$ be the leftmost variable in t.

LEMMA 7.1 ([8]). *The following identities hold in* $P = \langle \{1, 2, \ldots, n\}; f \rangle$

(1) $f(f, t_2, \ldots, t_k) \approx f$ *for any* k-*ary* $t_2, \ldots, t_k \in [f]$;

(2) $f(x_1, t_2, \ldots, t_k) \approx f$ *for any* k-*ary* $t_2, \ldots, t_k \in [f]$ *with* $\sigma(t_i) = x_i$ *for* $2 \leq i \leq k$;

(3) $f(x_1, t_2, \ldots, t_k) \approx x_1$ *for all* $t_2, \ldots, t_k \in [f]$ *provided* $\sigma(t_i) = x_1$ *for some* i;

(4) $f(x_1, t_2, \ldots, t_k) \approx x_1$ *for any* $t_2, \ldots, t_k \in [f]$ *provided* $\sigma(t_i) = \sigma(t_j)$ *for some* $i \leq j$.

The proof that $[f]$ is minimal is by induction on the length of $t \in [f]$ and uses (1)–(4) to show that, if t is not a projection, then an identification and permutation of variables reduces t to f.

Compare these clones with those minimal clones described in Theorem 6.8: define $A = \langle \{1, \ldots, n\}; g \rangle$ by $g(a_1, \ldots, a_k) = k$ if $a_1 = 1$, $a_k = k$ and $\{a_2, \ldots, a_{k-1}\} = \{2, \ldots, k-1\}$, and otherwise a_1. These algebras look quite similar. Yet, neither is in the variety generated by the other: $f(x_1, \ldots, x_{k-1}, f) \approx x_1$ while $g(x_1, \ldots, x_{k-1}, g) \approx g$. On the other hand, any absorption identity true in A is true in P.

QUESTION 7.2. Is it possible to make a strong link between minimal semiprojection clones and minimal conservative semiprojection clones via absorption identities?

REFERENCES

[1] BUEKENHOUT, F., DELANDTSHEER, A., DOYEN, J., KLEIDMAN, P., LIEBECK, M. and SAXL, J., *Linear spaces with flag transitive automorphisms groups.* Geom. Dedicata *36* (1990), 89–94.

[2] CSÁKÁNY, B., *All minimal clones on a three-element set.* Acta Cybernet. *6* (1983), 227–238.

[3] CSÁKÁNY, B., *On conservative minimal operations.* [Coll. Math. Soc. János Bolyai, No. 43]. North-Holland, Amsterdam, 1986, 49–60.

[4] JEŽEK, J. and QUACKENBUSH, R., *Directoids: algebraic models of up-directed sets.* Algebra Univ. *27* (1990), 49–69.

[5] JEŽEK, J. and QUACKENBUSH, R., *Minimal clones of conservative functions.* Preprint.

[6] MCKENZIE, R., *Para primal varieties: A study of finite axiomatizability and definable principal congruences in locally finite varieties.* Algebra Univ. *8* (1978), 336–348.

[7] PADMANABHAN, R. and WOLK, B., *Equational theories with a minority polynomial.* Proc. Amer. Math. Soc. *83* (1981), 238–242.

[8] PALFY, P. P., *The arity of minimal clones.* Acta Sci. Math. (Szeged) *50* (1986), 331–333.

[9] PALFY, P. P., *Minimal clones.* Manuscript.

[10] PÖSCHEL, R. and KAULUŽNIN, L. A., *Funktionen- und Relationenalgebren.* Deutscher Verl. Wiss., Berlin, 1979.

[11] ROSENBERG, I. G., *Minimal clones I: the five types.* [Coll. Math. Soc. János Bolyai, No. 43]. North-Holland, Amsterdam, 1986, 405–427.

[12] ŚWIERCZKOWSKI, S., *Algebras which are independently generated by every n elements*. Fund. Math. *49* (1960), 93–104.

[13] SZENDREI, Á, *Clones in universal algebra*. [Sém. de Math. Supp., No. 99]. Les Presses de l'Université de Montréal, Montréal, 1986.

[14] SZENDREI, Á, *Idempotent algebras with restrictions on subalgebras*. Acta Sci. Math. (Szeged) *51* (1987), 251–268.

Department of Mathematics,
University of Manitoba,
Winnipeg, Manitoba R3T 2N2,
Canada.

Aequationes Mathematicae **50** (1995) 17–37
University of Waterloo

0001–9054/95/020017–21 $1.50 + 0.20/0

General theory of the translation equation

ZENON MOSZNER

Summary. This paper gives a survey of the results of the general theory of translation equation which appeared after 1973.

Introduction

In [41] I listed several mathematical domains (abstract geometric and algebraic objects, abstract automata, groups of transformations, iterations, linear representations of groups, dynamical systems) in which the translation equation appears. I showed there also some directions of the development of the general theory of this equation. The bibliography in [41] contains papers in these directions published until 1973. So, in this survey we consider papers concerning the general theory of the translation equation which appeared after 1973 (both in the directions mentioned in [41] and in new domains of this theory).

The equation

$$F(F(\alpha, x), y) = F(\alpha, x \cdot y), \tag{1}$$

where the unknown function F has its values in an arbitrary set Γ and is defined on a subset of the Cartesian product $\Gamma \times G$, where G is a set with a binary operation " $\cdot$ ", defined for some pairs $(x, y) \in G \times G$, is called the translation equation (or the transformation equation).

If the function F is defined on the whole set $\Gamma \times G$ and the operation " $\cdot$ " is defined for all pairs $(x, y) \in G \times G$, then the notion that F satisfies the translation equation needs no comments: both sides of the equation are defined and equal for

---

AMS (1991) subject classification: Primary 39B50, Secondary 39B40.

Manuscript received October 6, 1992 and, in final form, March 30, 1993.

any $\alpha \in \Gamma$ and $x, y \in G$. The situation changes when we deal with the general case, as it has been formulated above. One can define in various ways what it means that the translation equation is satisfied, and in various fields of mathematics we do use various definitions.

To this area belong the papers [27], [62], [82], [83], from which I quoted the first two in [41] with incomplete data. They give a complete system of definitions stating when the translation equation is satisfied, comparison of these definitions and their consequences, e.g. for the problem of extending solutions.

I. Structure of solutions

The papers [60], [76], [65], [63], [64], [15], [80], [14], [5], [24], contain constructions of the general solution of the translation equation on some algebraic structures. No additional assumptions are supposed. I present here the result of one of these papers. The paper [80] gives some simplification of the construction of the general solution of (1) in the case where G is a group (see [41]). It reads as follows.

Let $\{G_k\}_{k \in K}$ denote an arbitrary family of subgroups of the group G. We do not assume that the map $k \to G_k$ is one-to-one, thus the map $k \to G/G_k$, where $G/G_k = \{G_k x : x \in G\}$, is also not necessarily one-to-one. We shall introduce the so called indexed quotient structure $(G/G_k, k) := \{(G_k x, k) : x \in G\}$ for $k \in K$. In this way we shall obtain a one-to-one map $k \to (G/G_k, k)$.

The multiplication of indexed cosets by elements of the group G is defined in the natural way:

$$(G_k x, k) y := (G_k xy, k) \qquad \text{for } x, y \in G, k \in K.$$

We choose

(1) an arbitrary family $\{G_k\}_{k \in K}$ of subgroups of G such that

$$\operatorname{card} \bigcup_{k \in K} (G/G_k, k) \leq \operatorname{card} \Gamma;$$

(2) an arbitrary one-to-one map $\phi \colon \bigcup_{k \in K} (G/G_k, k) \to \Gamma$ and
(3) an arbitrary function $g \colon \Gamma \to \Gamma$ such that $g(g) = g$ and

$$\phi\left(\bigcup_{k \in K} (G/G_k, k) \right) = g(\Gamma).$$

We define the function F by the formula

$$F(\alpha, x) := \phi[\phi^{-1}(g(\alpha))x] \qquad \text{for } \alpha \in \Gamma, x \in G. \tag{2}$$

If we know all solutions of (1) on a structure G, it was shown in [60] how to get all solution of this equation on the structure obtained by adjoining a zero 0 (i.e. we set $0 \cdot x = x \cdot 0$ for $x \in G \cup \{0\}$, where $0 \notin G$).

In Chapter VIII of [76] the general structure of the solution of (1) on $G \cup \{0\}$ was given in the case where G is an Ehresmann groupoid.

In [63] the general solution of (1) was given in the case where $\operatorname{card}(G \cdot G) = 1$ or $ab = b$ for all $a, b \in G$ or $ab = a$ for all $a, b \in G$. The paper [5] generalizes these results to the case of equation

$$F(\ldots F(F(\alpha, x_1), x_2), \ldots), x_n) = F(\alpha, x_1 \cdot x_2 \ldots x_n).$$

Under the assumption that $x_1 \cdot x_2 \ldots x_n = x_n$ for all $x_1, \ldots, x_n \in G$ we have the following result.

A mapping $F \colon \Gamma \times G \to \Gamma$ is a solution of the equation

$$F(\ldots F(F(\alpha, x_1), x_2) \ldots), x_n) = F(\alpha, x_n)$$

iff there exist a partition $\{\Gamma_i\}_{i \in I}$ of Γ, a family J of functions $f \colon \Gamma \to \Gamma$ and a function $h \colon G \to J$ such that the following conditions are satisfied:

(1) if $f \in J$ then $f^{n-1}(\Gamma_i) \subset \Gamma_i$ and $\operatorname{card} f(\Gamma_i) = 1$ for every $i \in I$;
(2) $f_1(f_2) = f_1(f_3)$ for all $f_1, f_2, f_3 \in J$:
(3) $F(\alpha, x) = (h(x))(\alpha)$ for all $(\alpha, x) \in \Gamma \times G$.

In [14] the construction of the general solution of (1) was given in the case where G is a commutative semigroup, $G \cdot G = G$ and the order $\leq$ defined by $x \leq y \leftrightarrow x + y = y$ is linear, complete and G possesses a minimal element.

The long paper [24] concerns, among other things, the construction of the general solution of (1) on some not necessarily associative structures called QD-groupoids.

A groupoid $(G, \cdot)$ is called a groupoid with quasi-division (QD-groupoid), if the condition

$$\forall a, b \in G \; \exists c, d \in G \qquad (\overline{a, c, b} = a \text{ or } \overline{a, d, b} = b)$$

is satisfied, where, by definition,

$\overline{a, c, b}$ equals either $a(cb)$ or $(ac)b$ $(a, b, c \in G)$.

Obviously, $\overline{a, c, b}$ is uniquely defined iff $(G, \cdot)$ is a semigroup.

It is easy to verify that every groupoid with division (D-groupoid) is a QD-groupoid. Consequently, quasigroups, loops and groups are QD-groupoids.

For QD-groupoids the solution of the translation equation can be characterized by means of partitions P of the set G satisfying the condition

$$\forall a \in G \ \forall A \in P \ \exists B \in P \quad (A \cdot a \subset B), \tag{3}$$

and the following condition

$$(Aa)b \subset B \rightarrow A(ab) \subset B \quad \text{for } a, b \in G; A, B \in P.$$

Such partitions are called the translative partitions of $(G, \cdot)$.

In [25] it is proved that, in the case where $(G, \cdot)$ is a quasigroup, every solution of (1) can be characterized by means of the translative partitions of $(G, \cdot)$. Moreover, it has been shown that each translative partition of $(G, \cdot)$ is of the form $\{Ha: a \in G\}$, where H is some special subquasigroup of $(G, \cdot)$, called translative subquasigroup of $(G, \cdot)$. In the paper [24] we generalize this result. In the class of QD-groupoids we characterize the solution of equation (1) by means of the translative partitions, and we characterize also the translative partitions by means of some special QD-groupoids in $(G, \cdot)$, called translative QD-groupoids in $(G, \cdot)$. In this case the main problem is not the form of solutions of (1) but the characterization of the translative partitions. This is related to the fact that we do not assume associativity and that there need not exist local identities and inverses. It compels to distinguish some elements, the role of which is similar to that of local identities and inverses. The translative QD-groupoids play for the QD-groupoids an analogous role to that of the subgroups for groups.

Several more papers were published in the considered period which give only some classes of the solutions of equation (1), for instance [6], [7], [42], [15], [39], [33], [50], [54], [2].

The papers [6], [7], [39] generalize results of [4]. For example, in [39] the following theorem was proved.

If

(1) Γ *is a Hausdorff topological space and a function* $F: \Gamma \times R \rightarrow \Gamma$ *satisfies the translation equation*

$$F(F(\alpha, x), y) = F(\alpha, x + y),$$

(2) *for some $\alpha_0 \in \Gamma$ the function $g(y) = F(\alpha_0, y)$ is continuous and not constant,*

(3) *F is transitive, i.e.*

$$\forall \alpha, \beta \in \Gamma \; \exists x \in R \qquad F(\alpha, x) = \beta,$$

then the function $g: R \to \Gamma$ is a bijection such that

$$F(\alpha, x) = g[g^{-1}(\alpha) + x]$$

on the set $\Gamma \times R$ or there exists a constant $c > 0$ such that the function $g_{\|[0,c)}: [0, c) \to \Gamma$ is a bijection with which

$$F(\alpha, x) = g_{\|[0,c)}\left[g_{\|[0,c)}^{-1}(\alpha) \underset{\text{mod } c}{+} c \right]$$

on the set $\Gamma \times R$ (here $\underset{\text{mod } c}{+}$ denotes the addition modulo c).

In [42] all the solutions $F: \Gamma \times G^+ \to \Gamma$ of equation (1) were given, where Γ is an arbitrary set and G^+ is a semigroup of positive elements of an Archimedean group G, satisfying the condition

$$\forall \alpha, \beta \in F(\Gamma, G^+) \qquad [F(\alpha, G^+) \cap F(\beta, G^+) \neq \varnothing$$

$$\Rightarrow F(\alpha, G^+) \subset F(\beta, G^+) \text{ or } F(\beta, G^+) \subset F(\alpha, G^+)]. \tag{4}$$

These solutions are given by the following *construction (C)*:

(1) Let $f: \Gamma \to \Gamma$ be a mapping satisfying the condition

$$\forall \alpha \in \Gamma: f(f(\alpha)) = f(\alpha).$$

(2) We decompose $f(\Gamma)$ into a disjoint union of non-empty sets Γ_k ($k \in K$) such that for each $k \in K$ there exists an invariant decomposition $\{W_{ik}\}_{i \in I_k}$ (i.e. satisfying (3) for $P = \{W_{ik}\}_{i \in I_k}$) of the interval $\Delta_k \subset G$ such that $G^+ \subset \Delta_k$ and card $I_k =$ card Γ_k.

(3) Let $h_k: \{W_{ik}\}_{i \in I_k} \to \Gamma_k$ be a bijection and let us define $h_k^*: \Delta_k \to \Gamma_k$ in the following way

$$h^*(x) = h_k(W_{ik}) \qquad \text{for } x \in W_{ik}.$$

(4) We put

$$F(\alpha, x) = h_k^*(h_k^{*-1}(f(\alpha))x) \qquad \text{for } f(\alpha) \in \Gamma_k.$$

The decompositions mentioned above were given in [43], [32], [35].

Some generalizations of the construction (C) were given in [15] and [33].

The papers [50], [54], [2] generalize, among other, the particular form of the solution of (1), known from [3], [1], to the following:

$$F(\alpha, x) = f^{-1}(f(\alpha)l(x)), \tag{a}$$

where f is a bijection of Γ onto group G_1 isomorphic to the group G, l is a homomorphism of G into G_1,

$$F(\alpha, x_1, \ldots, x_m) = f^{-1}(f(\alpha) + c_1 x_1 + \cdots + c_m x_m), \tag{b}$$

where f is a homomorphism of Γ into R and $(c_1, \ldots, c_m) \in R^m$,

$$F(\alpha, x) = k^{-1}(k(\alpha) + Cx), \tag{c}$$

where $F: R^n \times R^m \to R^n$, $k: R^n \to R^n$ is a bijection and C is a constant $n \times m$ matrix of rank $\min(n, m)$,

$$F(\alpha, x) = \Phi^{-1}(\Phi(\alpha) + c(x)), \tag{d}$$

where $F: \Gamma \times G \to \Gamma$, Φ is an injection of Γ into a group H, c is a homomorphism of the group G into H,

$$F(\alpha, x) = f^{-1}[(u + vB(u))J_\varrho + f(u)], \tag{e}$$

where $F: R^n \times R^m \to R^n$, f is a bijection of R^n onto itself, $\varrho = \min(n, m)$, $x = (x_1, \ldots, x_\varrho, \ldots, x_n)$, $u = (x_1, \ldots, x_\varrho)$, $v = (x_{\varrho+1}, \ldots, x_n)$, $B(\alpha)$ for α from R^n is a $(m - \varrho) \times \varrho$ matrix satisfying

$$B(F(\alpha, x)) = \beta(\alpha),$$

$J_\varrho = (1_\varrho, 0)$, 1_ϱ is a unit $\varrho \times \varrho$ matrix and 0 denotes $\varrho \times (n - \varrho)$ zero matrix.

Moreover, in these papers necessary and sufficient conditions are given for a solution of (1) to be of one of the forms given above.

The form (e) of the solution of (1) and the problem, considered in [2], when the composition of two solutions of form (e) is of this form as well, generated the papers [9], [10], [11] in which these problems are considered locally. In [21], the more general translation equation of "Pexider's type"

$$F_1(F_2(\alpha, x), y) = F_3(\alpha, xy),$$

is considered, where the unknown functions F_1, F_2, F_3 are defined on subsets of a set $\Gamma \times E$, where Γ is an arbitrary set, E is in particular an Ehresmann groupoid [78] and F_1, F_2, F_3 take their values in Γ.

E. J. Jasińska and M. Kucharzewski [28], [29] began to define accurately the notion of the Klein geometry. These considerations and their continuation in [72] led to an interesting possibility of expressing a solution of the translation equation in terms of a particular solution. Generally this result can be formulated as follows [78].

Given a solution $f: \Gamma_1 \times G \to \Gamma_1$ of (1), where G is a group and e its unit, satisfying the identity condition

$$\forall \alpha \in \Gamma: f(\alpha, e) = \alpha,$$

and the effectivity condition

$$\forall x \in G \qquad [\forall \alpha \in \Gamma: f(\alpha, x) = \alpha \to x = e],$$

we can express with aid of f every solution $F: \Gamma_2 \times G \to \Gamma_2$ of equation (1) satisfying the identity condition if

$$\text{card } \Gamma_2 < \sup(\text{card } \Gamma_1, \text{card } 2^{\Gamma_1}, \text{card } 2^{2^{\Gamma_1}}, \dots).$$

Considerable applications of the general theory of equation (1) to the Klein geometry are possible (see e.g. [28], [29] and references in [41] and [76], [52]).

II. Regular solutions (continuous, differentiable, analytic, monotonic)

Continuous solutions (with respect to one variable or to both variables) of (1) appear natural in different applications of this equation, in particular to iteration theory. Therefore this domain contains many interesting papers concerning continuous solutions (e.g. by J. Aczél, J. A. Baker, C. Blanton, D. Gronau, E. Jabotinsky, H. Michel, M. Sablik, A. Sklar, G. Targoński, J. Weitkämper, M. C. Zdun) in which variables run over subsets of the set of real numbers with natural topology.

Since results of iteration theory is discussed in another survey, by G. Targonski, in this issue, I omit them in the rest of my survey and in the bibliography. Similarly, the theory of dynamical systems is a theory of regular solutions of (1) with some additional properties. It is impossible to discuss here the results of this theory, even those achieved in the last twenty years.

Therefore we present only the papers [46], [18], [17], [19], [55] briefly.

I quote from [17] the following theorem concerning differentiable solutions of equation (1), where Γ is a real interval and $(G, \cdot) = (R, +)$.

Let F be a solution of (1) *on* $\Gamma \times R$ *(where Γ is a real interval) which is differentiable with respect to each coordinate. Then F is of the structure*

$$F(\alpha, x) = \begin{cases} h_m(h_m^{-1}(\alpha) + x) & \text{when } \alpha \in J_m, \\ \alpha & \text{otherwise,} \end{cases}$$

where $\{J_m\}_{m \in M}$ are the components of an open set O, $O = \bigcup_{m \in M} J_m$, $h_m : R \to J_m$ is a differentiable homeomorphism from R onto J_m such that $h'_m(u) \neq 0$ for all $u \in R$ and for each $m \in M$.

Conversely, any function F of the above structure is a solution of (1), *differentiable with respect to the second variable. It is differentiable with respect to the first variable iff each of the functions $\alpha \to F(\alpha, x)$ is differentiable at any point $\alpha \in \delta O$, the border of O.*

Further, the author finds conditions on O or on the functions h_m which make F differentiable with respect to the first coordinate at the points $\alpha \in \delta O$.

The paper [19] deals with locally differentiable solutions of (1) in a Banach space, defined in the following way.

Let Γ be a real or complex Banach space. A Γ-valued function F is said to be a local solution of the equation (1) if there exist open neighbourhoods U, U' in Γ of $0 \in \Gamma$ and real (open or half open) intervals I, I' or open connected subsets I, I' of complex numbers, each containing the number 0 and where U' is a subset of U, I' a subset of I, such that F is defined on $U \times I$ and

$$F(\alpha, x) \in U \qquad \text{for } (\alpha, x) \in U' \times I'$$

and

$$F(F(\alpha, x), y) = F(\alpha, x + y)$$

holds for all $x, y \in I$ with $x + y \in I$ and $\alpha \in U$, whenever $F(F(\alpha, x), y)$ is defined.

The following theorem was proved there:

Let F be a solution of (1) *(in the sense of the above definition) with* $F(0, 0) = 0$ *and let* $F(\alpha, 0)$ *be continuously differentiable near zero. Then there exist closed subspaces* Y *and* K *of* Γ *and a local diffeomorphism* T *with fixed point zero, such that* $\Gamma = Y \oplus K$.

The conjugate

$$F^*(\alpha, x) = T(F(T^{-1}(\alpha), x)),$$

which is again a solution of (1), defined on some neighbourhood $W \times I$ of $(0, 0)$ in $\Gamma \times I$, satisfies

(i) $F^*(\beta + \gamma, 0) = \beta$ for $\beta + \gamma \in Y \oplus K$;

(ii) Im F^* is contained in Y;

(iii) $F^*(\beta + \gamma_j x) = F^*(\beta, x)$ for $\beta \in Y \cap W, \gamma \in K \cap W$, $x \in I$.

If $F(\alpha, x)$ is continuously differentiable with respect to the variable α for all x, then so is $F^*(\alpha, x)$. And if, with $x \in I$, also $-x$ is contained in I, then $F^*(\alpha, x)$ restricted to Y is a local diffeomorphism on Y with its inverse $F^*(\alpha, -x)$.

The paper [18] concerns monotonic and continuous solutions of (1), where Γ is a linearly ordered set and G a linearly ordered group. We say that a mapping $F: \Gamma \times G \to \Gamma$ is monotonic if

(1) for each $x \in G$ the mapping $F(\cdot, x)$ is monotonic (in the same direction) and

(2) for each $\alpha \in \Gamma$ the mapping $F(\alpha, \cdot)$ is monotonic (in the same direction).

As we know [41] that the general transitive (i.e. $\forall \alpha, \beta \in \Gamma \exists x \in G; F(\alpha, x) = \beta$) solution of (1) is given by the formula

$$F(\alpha, x) = g(g^{-1}(\alpha)x), \tag{5}$$

where g is a bijection of G/G^*—the set of right cosets of G modulo some of its subgroup G^*, onto Γ.

One can prove easily that, if G^* is a convex subgroup of G, then G/G^* is linearly ordered by the following relation

$$A \leq B \leftrightarrow \exists x \in A \ \exists y \in B: x \leq y \qquad \text{for } A, B \in G/G^*.$$

The general transitive and monotic solution of (1) is given by (5), where it is additionally required that G^* be convex and g be monotonic.

Assume now that Γ and G are endowed with topologies induced by the orders in Γ and G, respectively. Then every transitive and monotonic solution of (1) is continuous.

A characterization of continuous solutions of (1) in the general case, where Γ is a topological space and G a topological structure, is very difficult even in the case where G is a topological group and F is a non-transitive solution. These difficulties are considered in [46]. In that paper a condition is given under which continuity of a (not necessarily transitive) solution of (1) on a topological group is equivalent to the continuity of the parameters used in the construction of the solution.

At the University of Graz (D. Gronau, G. H. Mehring, L. Reich, J. Schwaiger) the theory of formal power series has been developed in which a condition called also the translation equation appears, however it is not an equation of the type (1) (see [55]). For this reason I am not going to present the interesting results of this group.

In [55] analytic solutions of (1) are considered, which are defined as follows:

Let $\{F(\alpha, x)\}_{x \in K}$, where $K = R$ or $K = C$, be a family of functions $F(\cdot, x): C(x) \to K$ for all $x \in K$, where $C(x) = \{\alpha \in K: |\alpha| < \varrho(x)\}$ for $\varrho(x)$ being real positive ($\varrho(x) = \infty$ is not excluded for some x), analytic with respect to α and such that

$$\forall x \in K: F(0, x) = 0,$$

i.e., such that

$$F(\alpha, x) = a_1(x)\alpha + a_2(x)\alpha^2 + \cdots$$

for each $x \in K$ and each $\alpha \in C(x)$, where $a_v(x): K \to K$.

The family $\{F(\alpha, x)\}_{x \in K}$ is an analytic solution of (1) provided that there exists an open interval I, symmetric with respect to zero or an open ball I with the centre at zero ($I = K$ is also possible) such that (1) is valid for all $\alpha \in I$ and $x, y \in K$.

In [55], among other things, it is proved that $F(\alpha, x) = a_1(x)\alpha$, where $a_1(x)$ is an exponential function, is the unique analytic solution of (1) in the case where $I = K$ or if

$$K = C \quad \text{and} \quad |a_1(x)| \neq 1 \quad \text{or} \quad a_1(x) = 1 \quad \text{or} \quad F(I, C) \subset I.$$

It has also been shown there that in the case $K = C$ the function

$$F(\alpha, x) = \frac{1}{2} e^{irex} \left[2\alpha\, e^{irex} + (1 - e^{irex}) \left(\begin{pmatrix} \frac{1}{2} \\ 1 \end{pmatrix} 4\alpha + \begin{pmatrix} \frac{1}{2} \\ 2 \end{pmatrix} (4\alpha)^2 + \cdots \right) \right]$$

$$= \frac{1}{2} e^{irex} [e^{irex} - 1 + 2\alpha\, e^{irex} + (1 - e^{irex})(1 + 4\alpha)^{1/2}]$$

is an analytic solution of (1), which is not of the form $a_1(x)\alpha$.

At the University of Innsbruck (W. Förg-Rob, K. Kuhnert, R. Liedl, H. Reitberger) a theory has been developed, by the method of the so-called Pilgerschritt transformation, which enables to determine one-parameter subgroups of a topological group and which has some applications to characterizations of regular solutions of the translation equation. The discussion of the results of this theory and of papers related to it (see, for example, the bibliography in [31]) is beyond the framework of this topic.

III. The problem of extension

For a given set Γ^* such that $\Gamma \subset \Gamma^*$ and a given structure G^*, such that G is a substructure of G^*, the problem arises of extending a solution $F: \Gamma \times G \to \Gamma$ of equation (1) to a solution $F^*: \Gamma^* \times G^* \to \Gamma^*$ of this equation. Different modifications of this problem can also be considered (see [44]). To this area belong the papers [27], [59], [74], [75], [76], [58], [64], [83], [13], [44], [47], [16], [38], [84], [34]. Below I give some results of some of these papers.

In [59] the following theorem was proved about the extendability in the case where $\Gamma^* = \Gamma$, G and G^* are groups and we do not assume the transitivity of F.

A function of the form (2) *is extendable from the set* $\Gamma \times G$ *to the set* $\Gamma \times G^*$ *iff*

$$\forall G_k \ \exists G_k^* \subset G^* \qquad (G_k^* \cap G = G_k)$$

and there exists a decomposition $\{K_l\}_{l \in L}$ *of the set* K *such that*

$$\forall l \in L \ \exists q \in K_l \ \forall k \in K_l \ \forall a_k \in G^* \ \left[G_k = a_k^{-1} G_q a_k \text{ and the family} \right.$$

$$\left. \{A_p\}_{p \in K_l}, \text{ where } A_p = \bigcup_{b \in G} G_q^* a_k b \text{ is a decomposition of } G^* \right].$$

Let A be an arbitrary set and G an arbitrary group. Consider the translation equation

$$F[F(\alpha; (a, b, x)), (b, c, y)] = F(\alpha; a, c, xy)) \tag{6}$$

for all $a, b, c \in A$, $\alpha \in \Gamma$ and all $x, y \in G$. A binary inner operation in $A \times A \times G$ is also defined as follows:

$$(a, b, x) \cdot (d, c, y) = (a, c, xy) \leftrightarrow b = d$$

(from the theory of geometric objects).

The general solution of this equation can be obtained in the following way [74], [76]:

(1) for every $a \in A$ we choose arbitrarily sets Γ_a and Γ_a^* such that $\Gamma_a^* \subset \Gamma_a \subset \Gamma$ and card $\Gamma_a = $ card Γ_a^*,

(2) for each $a \in A$ we construct a function f_a mapping Γ_a onto Γ_a^* such that $f_a(\alpha) = \alpha$ for $\alpha \in \Gamma_a^*$,

(3) for a fixed element $a_0 \in A$ and for every $a \in A$ we construct a bijection h_a of the set Γ_a^* onto the set $\Gamma_{a_0}^*$,

(4) we choose an arbitrary function H satisfying the translation equation on $\Gamma_{a_0}^* \times G$,

(5) we put

$$F[\alpha; (a, b, x)] = h_a^{-1} H[h_b f_b(\alpha), x] \qquad \text{for } \alpha \in \Gamma_b. \tag{7}$$

The following theorem is valid [74]:

Let F satisfy equation (6) and let A^* be a superset of A (i.e. $A \subset A^*$) and G^* a supergroup of G. The function F can be extended to a solution of equation (6) on the set $\Gamma \times A^* \times A^* \times G^*$ iff there exists a representation of F in form (7) such that the function H can be extended to a function $H^*: \Gamma_{a_0}^* \times G^* \to \Gamma_{a_0}^*$ satisfying equation (1) on the set $\Gamma_{a_0}^* \times G^*$.

Extensions of solutions of equation (1) on an Ehresmann groupoid are discussed also in chapter VIII of [76].

A different question is the problem of extendability of regular (e.g. continuous, open, differentiable) solutions of equation (1). I quote from [75] the following results.

If G is an open subgroup of a topological group G^* and Γ is a topological space then every extension on the set $\Gamma \times G^*$ of a transitive regular solution F of (1) (i.e. F is continuous and for each $\alpha \in \Gamma$ the mapping $x \to F(\alpha, x)$ is open), is transitive and regular, too.

If Γ is a topological T_1-space, G an algebraic subgroup of a topological group G^* and F a transitive, regular solution of (1) on $\Gamma \times G$, which is regularly extendable on the set $\Gamma \times G^*$, then, for every so called stability subgroup $G_\alpha := \{x \in G: F(\alpha, x) = \alpha\}$ of the solution F, there exists a topological subgroup G_α^* of G^* such that

$$G^* = G_\alpha^* \cdot G \quad \text{and} \quad G_\alpha = G_\alpha^* \cap G.$$

The following theorem was proved in [47].

If P is semigroup of the positive elements of a linearly ordered Archimedean group G and $F: \Gamma \times P \to \Gamma$ is a solution of (1) then F is extendable to a solution

$F^*\colon \Gamma \times G \to \Gamma$ *of* (1) *iff*

(A) *for an arbitrary α from Γ the cardinality of the set $E_\alpha(b) := \{x \in P\colon F(\alpha, x) = \beta\}$ does not depend on β from $F(\alpha, P)$,*

(B) *the relation R defined on the set $F(\Gamma, e)$ in the following way*

$$\alpha R \beta \leftrightarrow \exists x \in P[F(\alpha, x) = \beta \text{ or } F(\beta, x) = \alpha]$$

is translative,

(C) $\forall x \in P\colon \quad F(\Gamma, e) \subset F(\Gamma, x)$.

From [58] we quote the following theorem:

The conjunction of the conditions (A) and (B) mentioned above is necessary and sufficient for the existence of a set $\Gamma^* \supset \Gamma$ and a solution $F^*\colon \Gamma^* \times G \to \Gamma^*$ of (1) which is an extension of F.

The following was proved in [38].

If P is a subsemigroup of the group G such that $G = P \cup P^{-1}$ and $F\colon \Gamma \times P \to \Gamma$ is a solution of (1) *then there exists a solution $F^*\colon \Gamma \times G \to \Gamma$ of* (1) *which is an extension of F iff for all x from P the function $F(\,\cdot\,, x)_{|F(\Gamma, e)}$ is a bijection onto $F(\Gamma, e)$.*

We quote also the following theorem from [34]:

If P is a subsemigroup of the group G such that $G = P \cup P^{-1}$ and $F\colon \Gamma \times P \to \Gamma$ is a solution of (1) *then there exists a set $\Gamma^* \supset \Gamma$ and a solution $F^*\colon \Gamma^* \times G \to \Gamma^*$ of* (1) *which is an extension of F iff for all x from P the function $F(\,\cdot\,, x)_{|F(\Gamma, e)}$ is an injection into $F(\Gamma, e)$.*

IV. Additional properties of solutions

Various applications of solutions of equation (1) require that they satisfy some additional properties on them. It turns out that possession of these properties by solutions, in the case where G is a group, can be characterized by the parameters $g(\alpha)$ and G_k (the so called stability subgroups of the solution) determining solutions of the form (2). I am going to give, following [56], definitions of some of these properties and after each of them an equivalent condition formulated in terms of the parameters $g(\alpha)$ and G_k.

(1) The identity condition: $\forall \alpha \in \Gamma\colon F(\alpha, e) = \alpha$,
    —it is equivalent to $g(\alpha) = \alpha$.

(2) The transitivity: $\forall \alpha, \beta \in \Gamma \quad \exists x \in G: F(\alpha, x) = \beta$,
—it is equivalent to $g(\alpha) = \alpha$ and card $K = 1$.

(3) The quasi-transitivity: $\forall \alpha, \beta \in \Gamma \quad \exists x \in G: F(\alpha, x) = \beta$ or $F(\beta, x) = \alpha$,
—it is equivalent to $g(\alpha) = \alpha$ and card $K = 1$.

(4) The simple transitivity: the transitivity and the injectivity of $F(\alpha, \cdot)$ for each $x \in G$,
—it is equivalent to $g(\alpha) = \alpha$ and card $K = 1$ and $G_k = \{e\}$.

(5) The injectivity of $F(\cdot, x)$ for each $x \in G$,
—it is equivalent to $g(\alpha) = \alpha$.

(6) The effectivity: $\forall x \in G \; [\forall \alpha \in \Gamma: F(\alpha, x) = \alpha \rightarrow x = e]$,
—it is equivalent to $N := \bigcap_{k \in K} \bigcap_{a \in G} a^{-1} G a = \{e\}$.

(7) G acts freely on Γ by F, i.e. $\forall x \in G \; [\exists \alpha \in \Gamma: F(\alpha, x) = \alpha \rightarrow x = e]$,
—it is equivalent to $G_k = \{e\}$ for each $k \in K$.

(8) F is disjoint at a point α_0, i.e.

$$\forall x, y \in G \quad [F(\alpha_0, x) = F(\alpha_0, y) \rightarrow \forall \alpha \in \Gamma: F(\alpha, x) = F(\alpha, y)].$$

—it is equivalent to $G_{k_0} = N$ for k_0 such that $g(\alpha_0) \in \Gamma_{k_0}$.

(9) The commutativity of F, i.e.

$$\forall x, y \in G \; \forall \alpha \in \Gamma: F(F(\alpha, x), y) = F(F(\alpha, y), x),$$

—it is equivalent to the condition that the quotient group G/N is abelian.

(10) F is maximal, i.e.

$$\forall p: \Gamma \rightarrow \Gamma \quad [\forall \alpha \in \Gamma \; \forall x \in G: p(F(\alpha, x)) = F(p(\alpha), x) \rightarrow$$

$$\exists x_0 \in G \; \forall \alpha \in \Gamma: p(\alpha) = F(\alpha, x_0)],$$

—it is equivalent to the following:

$$g(\alpha) = \alpha \quad \text{and} \quad G_k \neq G \text{ for each } k \in K$$

$$\text{and} \quad \bigcap_{k \in K} G_k a(k) \neq \emptyset \text{ for each function } a: K \rightarrow G.$$

(11) F is called parallelizable if there exists such a function $\tau: \Gamma \rightarrow G$ that
$\tau(F(\alpha, x)) = \tau(\alpha) x^{-1}$,
—it is equivalent to the following: G is abelian and $G_k = \{e\}$ for each $k \in K$.

The above equivalences were proved in the papers [60], [8], [51], [53], [56]. In the case where G forms a structure more general than a group, the conditions in these equivalences are more complicated. The commutativity of a solution, in the case where G is e.g. an Ehresmann groupoid, is discussed in [30]. The papers [73] and [77] refer to commutativity as well. For a structure with a zero element, in place of G, some of these conditions are discussed in papers [60] and [51] and for the case where G is a semigroup of positive elements of an Archimedean group in [51], [36], [37]. In [37] I give, as an example, a condition equivalent to the effectivity of a solution of (1) in the case where G^+ is a semigroup of positive elements of an Archimedean group G and the solution satisfies the condition (4).

Referring to the above mentioned construction (C) we give an arbitrary invariant decomposition of the interval Δ_k:

(i) there exists an element u from the complement of G such that points of the interval $J_k := \{x \in \Delta_k : x \ll u\}$, where $\ll = \leq$ or $\ll = <$, form components of this decomposition ($J_k = \varnothing$ is not excluded),

(ii) the remaining components of the decomposition are restrictions of cosets G/G_k to the set $\Delta_k \backslash J_k$, where G_k is a subgroup of G.

Then a condition equivalent to the effectivity of a solution of (1) is the following:

$$N = \bigcap_{k \in K} G_k = \{e\} \quad \text{or} \quad G^+ \subset \bigcup_{k \in K} \{xy^{-1} : x, y \in J_k\}.$$

V. Translation equation on the products of structures

In the papers [40], [30], [61], [12], [48] a solution of the translation equation on the direct product or on the semi-direct product of some structures is expressed by solutions of this equation on the components of the product. The results of the paper [40] have been generalized in [12] and [48].

VI. Homomorphisms and solutions of the translation equation

A function $F: \Gamma \times G \to \Gamma$ is a solution of equation (1) iff the mapping $x \to F(\cdot, x)$ of the structure $(G, \cdot)$ into the family of functions mapping Γ into itself (with the composition as a binary operation), is a homomorphism. So, every solution of (1) is a homomorphism. The converse is also true. Each homomorphism h of the structure G into an associative structure S dictates a solution F of equation (1), with S in place of the fibre Γ. This homomorphism is defined by the following condition: $F(\alpha, x) = \alpha h(x)$. This means that a correlation exists between the solutions of (1) and homomorphisms, which has its consequences in the theory of the translation equation. There are some papers establishing also these consequences. To this area belong the papers [20], [22], [23], [26], [13], [45], [49]. Good examples are the results of the paper [21], which are consequences of the theorems concerning homomorphisms, given in [20] and [13].

VII. Set-valued iteration semigroups

Let A, X, Y and Z be nonempty sets with $A \subset X$ and let F be a set-valued function from X into Y, i.e. the values of F are subsets of Y. The image of A under F is the set

$$F(A) = \bigcup \{F(x): x \in A\}.$$

Moreover, if G is a set-valued function from Y into Z then one can define the composition $G(F)$ of F and G:

$$(G(F))(x) := G(F(x)).$$

A family $\{F^t, t > 0\}$ of set-valued functions F^t from X into X is said to be an iteration semigroup if the equation

$$F^s(F^t) = F^{s+t} \tag{8}$$

holds for every $s, t > 0$ and sets $F^s(x) \neq \emptyset$ for every $s > 0, x \in X$.

Let (X, ϱ) be a separable metric space, then the set $c(X)$ of all nonempty compact subsets of X is a separable metric space with respect to the Hausdorff metric.

An iteration semigroup of set-valued functions $F^t: X \to c(X)$ is said to be measurable (continuous) if the set-valued functions

$$t \to F^t(x) \qquad (x \in X) \tag{9}$$

are measurable (continuous) with respect to the Hausdorff metric.

If $F^t: X \to c(X)$ is an iteration semigroup of set-valued functions and X is a locally compact space and $F^t(x) \subset g(t, x)$ for $t > 0$, $x \in X$, where g is an upper semi-continuous compact-valued function, or X is compact or the $F^t(x)$ are lipschitzian in x, then the measurability of F^t implies its continuity (see [67] and see also [79] for single-valued iteration semigroups).

We say that the family F' fulfils functional equation (8) almost everywhere (a.e.) if the set of all pairs (s, t) $(s, t > 0)$ for which equation (8) does not hold, is a set of Lebesgue measure zero.

Suppose that X is a nonempty and closed subset of a separable Banach space and assume that $F^t: X \to c(X)$ is a family of lipschitzian set-valued functions such that every function (9) is Lebesgue measurable for $x \in X$. If equation (8) is fulfilled for a.e. $s, t > 0$ in X, then there exists a continuous iteration semigroup $\{G^t, t > 0\}$ of lipschitzian set-valued functions on X such that $G^t = F^t$ a.e. on $(0, \infty)$ (see [69]).

Let X be a locally compact and separable metric space. If $F^t: X \to c(X)$ is an iteration semigroup of contractions such that the set-valued functions (9) are upper semi-continuous then there exists a minimal semi-continuous iteration semigroup $\{G^t, t > 0\}$ of contractions $G^t: X \to c(X)$ such that $G^t(x) \subset F^t(x)$ for $t > 0$ and $x \in X$. This iteration semigroup is continuous (see [68]).

Let X be a nonempty subset of a linear space and let $\phi: X \to R$. A set-valued function F from X into X is said to be ϕ-increasing if, for every $x, y \in X$ with $\phi(x) \leq \phi(y)$ and every $w \in F(y)$, there exists a $u \in F(x)$ such that $\phi(u) \leq \phi(w)$.

Let X be a non-empty convex subset of a normed linear space and let ϕ be a real strictly convex and lower semicontinuous function defined on X. If $\{F^t, t > 0\}$ is an iteration semigroup of ϕ-increasing set-valued functions from X into X with convex, compact values, then there exists an iteration semigroup $\{f^t, t > 0\}$ of single-valued functions from X into X such that $f^t(x) \in F^t(x)$ and $\phi(f^t(x)) = \inf\{\phi(y): y \in F^t(x)\}$ for every $x \in X$ and $t > 0$. If ϕ and the set-valued functions $x \to F^t(x)$ $(t \to F^t(x))$ are continuous then the functions $x \to f^t(x)$ $(t \to f^t(x))$ are continuous (see [67]).

We note that set-valued solutions of the generalized translation equation, analogous to (8), occur in the theory of abstract, nondeterministic automata (see [70] or [71]) as functions of passage.

VIII. Some open problems

(1) Comparison of various definitions of the local solution of the translation equation (e.g. in [11], [19], [55]) and establishing conditions for extendability of these local solutions to global ones (see [57]).

(2) Constructions of solutions of translation equation on various algebraic structures by means of independent parameters.

(3) Problem of stability of the translation equation.

Let (Γ, ϱ) be a metric space and let G be a group. Does there exist for each $\varepsilon > 0$ a $\delta > 0$ such that for each $H: \Gamma \times G \to \Gamma$ satisfying the condition

$$\forall \alpha \in \Gamma, \forall x, y \in G: \varrho(H(H(\alpha, x), y), H(\alpha, x \cdot y)) < \delta$$

there exists a solution F of the translation equation such that

$$\forall \alpha \in \Gamma \; \forall x \in G: \varrho(H(\alpha, x), F(\alpha, x)) < \varepsilon?$$

For the equation $F(F(\alpha)) = F(\alpha)$ the answer is positive.
Some open problems are formulated in [56].

Acknowledgement

I would like to thank Professor János Aczél for his valuable remarks.

REFERENCES

[1] ACZÉL, J., *Lectures on functional equations and their applications.* Academic Press, New York–London, 1966.

[2] ACZÉL, J., BERG, L. and MOSZNER, Z., *Sur l'équation de translation multidimensionnelle.* Results in Math. *19* (1991), 195–210.

[3] ACZÉL, J. and HOSSZÚ, M., *On transformation with several parameters and operations in multidimensional spaces.* Acta Math. Acad. Sci. Hungar. *6* (1956), 327–338.

[4] ACZÉL, J., KALMAR, L. and MIKUSIŃSKI, J. G., *Sur l'équation de translation.* Studia Math. *12* (1951), 112–116.

[5] ADAMASZEK, J., *Solution of the generalized translation equation on some structures.* Univ. Jagiell. Acta Math. *27* (1988), 51–60.

[6] BANAŚ, J., *Translation equation on additive semigroup of reals non-negatives numbers (Polish).* Wyż. Szkoła Ped. Rzeszów. Rocznik Nauk.-Dydakt. No. *3/22* (1975), 3–14.

[7] BANAŚ, J. and MIDURA, S., *Construction of the solution of the translation equation on additive semigroup of real numbers (Polish).* Wyż. Szkoła Ped. Rzeszów. Rocznik Nauk.-Dydakt. No. *4/32* (1977), 7–30.

[8] BARCZ, E, KANIA, M. and MOSZNER, Z., *Sur les automates commutatifs.* Zeszyty Nauk. Uniw. Jagielloń., Prace Mat. *19* (1977), 195–199.

[9] BERG, L., *Fünf mehrdimensionale Funktionalgleichungen.* Mitt. Math. Ges. Hamburg *12* (1991), 697–703.

[10] BERG, L., *The composition of solutions of the multidimensional translation equation.* Results in Math., in print.

[11] BERG, L., *The local structure of the solutions of the multidimensional translation equation.* Aequationes Math. *46* (1993), 164–173.

[12] BÖNISCH, K. C., *L'équation de translation sur des extensions des groupes.* Wyż. Szkoła Ped. Rzeszów. Rocznik Nauk. Dydakt. No. *6/50* (1982), 7–14.

[13] DANKIEWICZ, K. and MOSZNER, Z., *Prolongements des homomorphismes et des solutions de l'équation de translation.* Wyż. Szkoła Ped. Kraków. Rocznik Nauk.-Dydakt. No. 82, Prace Mat. *10* (1982),. 27–44.

[14] DABROWSKA, A., *Solution of the translation equation on some semi-group semilattice (Polish)* .Wyż. Szkoła Ped. Rzeszów. Rocznik Nauk.-Dydakt. No. *6/50* (1982), 15–22.

[15] DABROWSKA, A., *Translation equation on the semigroup of elements positives of the Archimedean group (Polish).* Wyż. Szkoła Ped. Rzeszów. Rocznik Nauk.-Dydakt. No. *6/50* (1982), 23–32.

[16] ETGENS, M. and MOSZNER, Z., *On pseudo-processes and their extensions.* In: *Iteration Theory*

and its Functional Equations (ed. R. Liedl et al.). [Lecture Notes in Math., Vol. 1165]. Springer Verlag, Berlin–Heidelberg–New York, 1985, 49–58.

[17] FÖRG-ROB, W., On differentiable solutions of the translation equation. Grazer Math. Berichte 287 (1988), 1–17.

[18] GREGORCZYK, A. and TABOR, J., Monotonic solution of the translation equation. Ann. Polon. Math. 43 (1983), 253–260.

[19] GRONAU, D., Some properties of the solutions of the translation equation in Banach spaces. Grazer Math. Berichte 290 (1988), 1–14.

[20] GRZĄŚLEWICZ, A., On the solution of the generalizing equation of homomorphism. Wyż. Szkoła Ped. Kraków. Rocznik Nauk.-Dydakt. No. 61 Prace Mat. 8 (1977), 31–60.

[21] GRZĄŚLEWICZ, A., On the solutions of the equation $F_1(F_2(x, \beta), \alpha) = F_3(x, \alpha\beta)$. Wyż. Szkoła Ped. Kraków. Rocznik Nauk.-Dydakt. No. 61 Prace Mat. 8 (1977), 61–78.

[22] GRZĄŚLEWICZ, A., On extensions of homomorphisms. Aequationes Math. 17 (1978), 199–207.

[23] GRZĄŚLEWICZ, A., On some homomorphisms in product Brandt groupoids. Wyż. Szkoła Ped. Kraków. Rocznik Nauk.-Dydakt. No. 69 Prace Mat. 9 (1979), 67–72.

[24] GRZĄŚLEWICZ, A., Translations and transitive partitions of QD-groupoids. [Etudes Monographiques de l'École Normale Supérieure à Cracovie, Vol. 114]. W.S.P., Kraków, 1992.

[25] GRZĄŚLEWICZ, A., Translations and translative partitions of quasigroups. Wyż. Szkoła Ped. Kraków. Rocznik Nauk.-Dydakt. No. 159 Prace Mat. 13 (1993), 125–159.

[26] GRZĄŚLEWICZ, A. and SIKORSKI, R., On some homomorphisms in Ehresmann groupoids. Wyż. Szkoła Ped. Kraków. Rocznik Nauk.-Dydakt. No. 69 Prace Mat. 9 (1979), 55–66.

[27] GRZĄŚLEWICZ, A. and TABOR, J., On the equivalence of two definitions of the translation equation and the extensions of the solutions of this equation. Wyż. Szkoła Ped. Kraków. Rocznik Nauk.-Dydakt. No. 51 Prace Mat. 7 (1974), 47–57.

[28] JASIŃSKA, E. J. and KUCHARZEWSKI, M., Kleinsche Geometrie und Theorie der geometrischen Obiekte. Colloq Math. 26 (1972), 271–279.

[29] JASIŃSKA, E. J. and KUCHARZEWSKI, M., Grundlegende Bergriffe der Kleinschen Geometrie. Demonstratio Math. 7 (1974), 391–402.

[30] KANIA-SIUDA, M., Sur la commutativité de la solution de l'équation de translation. Wyż. Szkoła Ped. Kraków. Rocznik Nauk.-Dydakt. No. 69 Prace Mat. 9 (1979), 73–83.

[31] LIEDL, R. and NETZER, N., Die Lösung der Translationsgleichung mittels schneller Pilgerschritttransformation. Grazer Math. Berichte 314 (1991), 1–54.

[32] MACH, A., Sur les décompositions invariantes du demi-groupe du groupe abélien linéairement ordonné. Wyż. Szkoła Ped. Kraków. Rocznik Nauk.-Dydakt. No. 69 Prace Mat. 11 (1985), 119–145.

[33] MACH, A., Some class of solutions of the translation equation (Polish). Kieleckie Studia Mat. 5 (1989), 23–55.

[34] MACH, A., On an extension of solution of the translation equation from a subsemigroup onto the group. Zeszyty Nauk. Wyż. Szkoła Ped.w Rzeszowie 2, Mat.-Fiz.-Tech. 1 (1990), 79–107.

[35] MACH, A., Invariant decompositions of a final interval of a linearly ordered and abelian group. Tensor (NS) 49 (1990), 1–8.

[36] MACH, A., Sur les solutions disjointes de l'équation de translation. Ann. Polon. Math. 52 (1991), 287–291.

[37] MACH, A., Sur l'effectivité des solutions de l'équation de translation. Zeszyty Nauk. Politech. Śląsk., Mat.-Fiz. 67 (1992), 123–129.

[38] MACH, A. and MOSZNER, Z., Sur les prolongements de la solution de l'équation de translation. Univ. Jagiell. Acta Math. 26 (1987), 53–61.

[39] MIDURA, S., Les solutions continues de l'équation de translation sur le groupe additif des nombres réels. Demonstratio Math. 19/3 (1986), 337–346.

[40] MIDURA, S. and TABOR, J., The translation equation on a direct product of groups. Ann. Polon. Math. 35 (1978), 223–228.

[41] MOSZNER, Z., *The translation equation and its application.* Demonstratio Math. *6/1* (1973), 309–327.

[42] MOSZNER, Z., *Solution générale de l'équation de translation sur un demi-groupe.* Wyż. Szkoła Ped. Kraków. Rocznik Nauk.-Dydakt. No. 69 Prace Mat. *9* (1979), 97–104.

[43] MOSZNER, Z., *Décompositions invariantes du demi-groupe des éléments non-négatifs du groupe archimédien.* Tensor (NS) *34* (1980), 8–10.

[44] MOSZNER, Z., *Le prolongement de la solution de l'équation et de l'inégalité de translation avec l'agrandissement borné de la fibre.* Zeszyty Nauk. Uniw. Jagielloń. Prace Mat. *23* (1982), 85–90.

[45] MOSZNER, Z., *Sur la continuité des homomorphismes.* Wyż. Szkoła Ped. Kraków. Rocznik Nauk.-Dydakt. No. 82 Prace Mat. *10* (1982), 97–100.

[46] MOSZNER, Z., *Quelques remarques sur les solutions continues de l'équation de translation sur un groupe.* Demonstratio Math. *15/1* (1982), 279–284.

[47] MOSZNER, Z., *Sur le prolongement de la solution de l'équation de translation.* Ann. Polon. Math. *40* (1983), 239–244.

[48] MOSZNER, Z., *Solution de l'équation de translation sur le produit simple des groupoïdes.* Zeszyty Nauk. AGH w Krakowie *57* (1984), 207–219.

[49] MOSZNER, Z., *Sur un problème au sujet des homomorphismes.* Aequationes Math. *32* (1987), 297–303.

[50] MOSZNER, Z., *Une généralisation d'un résultat de J. Aczél et M. Hosszú sur l'équation de translation.* Aequationes Math. *37* (1989), 267–278.

[51] MOSZNER, Z., *Sur les propriétés complémentaires des solutions de l'équation de translation.* Ann. Polon. Math. *52* (1990), 27–36.

[52] MOSZNER, Z., *Les objets abstraits comme les systèmes des sous-groupes.* Zeszyty Nauk. Politech. Śląsk., Mat.-Fiz. *64* (1991), 191–201.

[53] MOSZNER, Z., *Sur les solutions maximales de l'équation de translation.* Aequationes Math. *42* (1991), 154–165.

[54] MOSZNER, Z., *Sur une forme de la solution de l'équation de translation.* C.R. Math. Rep. Acad. Sci. Canada *13* (1991), 285–290.

[55] MOSZNER, Z., *Sur des itérations analytiques et des itérations formelles.* In: *Proceedings of ECIT 89, European Conference on Iteration Theory, Batschuns, Austria, 10–16 Sept. 1989* (ed. Ch. Mira et al.). World Scientific, Singapore, 1991, 257–271.

[56] MOSZNER, Z., *Les propriétés et les formes invariants des solutions de l'équation de translation.* Grazer Math. Berichte *316* (1992), 139–158.

[57] MOSZNER, Z., *Sur des solutions globales et des solutions locales de l'équation de translation.* C.R. Math. Rep. Acad. Sci. Canada *14/5* (1992), 219–224.

[58] MOSZNER, Z. and NOWAK, B., *A condition of an extension of the solution of the translation equation and its applications.* Wyż. Szkoła Ped. Rzeszów. Rocznik Nauk.-Dydakt. No. *5/41* (1979), 77–91.

[59] MOSZNER, Z. and PILECKA, B., *Sur le prolongement des objets géométriques non-transitifs.* Tensor (NS) *28* (1974), 63–66.

[60] MOSZNER, Z. and TABOR, J., *L'équation de translation sur une structure avec zéro.* Ann. Polon. Math. *31* (1976), 255–264.

[61] MOSZNER, Z., and WAŚKO, M., *L'équation de translation sur le produit simple des groupes avec zéro.* Zeszyty Nauk. Uniw. Jagielloń., Prace Mat. *20* (1979), 101–109.

[62] MOSZNER, Z., ŻUREK, M., *Sur les différentes définitions des solutions de l'équation de translation.* Wyż. Szkoła Ped. Kraków. Rocznik Nauk.-Dydakt. No. 51 Prace Mat. *7* (1974), 95–108.

[63] PIECHOWICZ, L. and SERAFIN, S., *Solution of the translation equation on some structures.* Zeszyty Nauk. Uniw. Jagielloń., Prace Mat. *21* (1979), 109–114.

[64] PILECKA, B., *Le prolongement de la solution de l'équation de translation.* Wyż. Szkoła Ped. Kraków. Rocznik Nauk.-Dydakt. No. 69 Prace Mat. *9* (1979), 105–118.

[65] SERAFIN, S., *Solution of translation equation on right group and on left group (Polish).* Wyż. Szkoła Ped. Rzeszów. Rocznik Nauk.-Dydakt. No. *5/41* (1979), 143–156.

[66] SERAFIN, S., *Solution of the translation equation on extensions of semigroup with zero-multiplication.* Wyż. Szkoła Ped. Kraków. Rocznik Nauk.-Dydakt. No. 82 Prace Mat. *10* (1982), 117–122.

[67] SMAJDOR, A., *Iterations of multi-valued functions.* Prace Nauk. Uniw. Śląsk. w Katowicach *759* (1985), 1–59.

[68] SMAJDOR, A., *One-parameter families of set-valued contractions.* In: *Proceedings of ECIT 87, European Conference on Iteration Theory, Caldes de Malavella, Spain, 20–26 Sept. 1987* (ed. C. Alsina et al.). World Scientific, Singapore, 1989, 324–330.

[69] SMAJDOR, A., *Almost-everywhere set-valued iteration semigroups.* In: *Proceedings of ECIT 91, European Conference on Iteration Theory, Lisboa, Portugal, 15–21 Sept. 1991* (ed. J. P. Lampreia et al.). World Scientific, Singapore, 1992, 262–272.

[70] STARKE, P. H., *Abstrakte Automaten.* Deutscher Verlag der Wissenschaften, Berlin, 1969.

[71] STARKE, P. H., *Abstract Automata.* North Holland Publishing Company, Amsterdam, 1972.

[72] SZOCIŃSKI, B., *Basic concepts of Klein geometries.* Zeszyty Nauk. Politech. Śląsk. 1055, Mat.-Fiz. *62* (1990), 1–81.

[73] TABOR, J., *Remarks on J. Gancarzewicz paper On commutative algebraic objects over a groupoid.* Zeszyty Nauk. Uniw. Jagielloń., Prace Mat. *17* (1975), 89–92.

[74] TABOR, J., *On the extension of algebraic objects over a Brandt groupoid.* Tensor (NS) *29* (1975), 85–88.

[75] TABOR, J., *Regular extensions of algebraic objects.* Tensor (NS) *30* (1976), 186–190.

[76] TABOR, J., *Algebraic objects over a small category.* Dissertationes Math. *155* (1978), 1–62.

[77] TABOR, J., *On commutative algebraic objects over the group GL(n, R).* Wyż. Szkoła Ped. Kraków. Rocznik Nauk.-Dydakt. No. 69 Prace Mat. *9* (1979), 159–163.

[78] TYSZKA, A., *On the notion of a geometric object in a Klein space.* Wyż. Szkoła Ped. Kraków. Rocznik Nauk.-Dydakt. No. 159 Prace Mat. *13* (1993), 287–299.

[79] WALISZEWSKI, W., *Categories, groupoids, pseudogroups and analytical structures.* Dissertationes Math. *45* (1965), 1–40.

[80] WOŁODŹKO, S., *A modification of a construction of Z. Moszner.* Wyż. Szkoła Ped. Kraków. Rocznik Nauk.-Dydakt. No. 82, Prace Mat. *10* (1982), 165–169.

[81] ZDUN, M. C., *Continuous and differentiable iteration semigroups.* Prace Nauk. Uniw. Śląsk. w Katowicach *308* (1979), 1–90.

[82] ŻUREK-ETGENS, M., *Sur une définition de remplissage de l'équation de translation.* Wyż. Szkoła Ped. Kraków. Rocznik Nauk.-Dydakt. No. 69 Prace Mat. *9* (1979), 191–193.

[83] ŻUREK-ETGENS, M., *La définition et le prolongement de la solution de l'équation de translation.* Demonstratio Math. *12/4* (1979), 889–902.

[84] ŻUREK-ETGENS, M., *On extension of a solution of the translation equation.* Wyż. Szkoła Ped. Kraków. Rocznik Nauk.-Dydakt. No. 115 Prace Mat. *12* (1987), 149–161.

Kazimierza Wielkiego 87/4,
PL-30-074 Kraków,
Poland.

Aequationes Mathematicae **50** (1995) 38–49
University of Waterloo

0001–9054/95/020038–12 $1.50 + 0.20/0

Some recent applications of functional equations to the social and behavioral sciences. Further problems

J. Aczél

Summary. Recent applications of functional equations to questions of allocation, aggregation, utility, taxation, theories of measurement and dimensional analysis are discussed and open problems formulated.

1. A surprisingly large number of recent applications are of the *aggregation* type. A (relatively) old example is "creating consensus in group decision making" (aggregation theorems for allocation problems, Table 1, see e.g. [1]. A relatively new one concerns equalization payments between two levels of government (for instance federal and states — Länder — provinces — cantons or between states etc. and municipalities; see e.g. [2], [3]. In its most recent incarnation the number of states n is supposed to be greater than two, but a fixed number. (That it has been considered variable previously is quite surprising. While changes in n can and do occur, in Germany as recently as 1990, this is still a rather singular occurrence and certainly then the equalization formula has to be revamped). For the k-th state we denote a hypothetical (planned or equidistributed) tax revenue by t_k, the actual (real) tax revenue by r_k, the budgeted federal subvention (grant) by s_k. The final total amount at its disposal will depend on these, equalling, say, $f_k(r_k, t_k, s_k)$ (so the actual federal equalization payment or subvention will be $f_k(r_k, t_k, s_k) - r_k$). As we know, the lower levels of government have their hands more tied even, or in particular, concerning taxation. In our model (Table 2) this is reflected so strongly

---

AMS (1991) subject classification: Primary 39B22, Secondary 90A07, 90A10, 90A70, 92G05.

Manuscript received December 3, 1992 and, in final form, August 23, 1993.

Table 1

Decision makers	Projects						Sums
	1	2	$\cdots$	k	$\cdots$	n	
1	x_{11}	x_{12}	$\cdots$	x_{1k}	$\cdots$	x_{1n}	C
$\vdots$	$\vdots$	$\vdots$		$\vdots\vdots\vdots$		$\vdots$	$\vdots$
j	x_{j1}	x_{j2}	$\cdots$	x_{jk}	$\cdots$	x_{jn}	C
$\vdots$	$\vdots$	$\vdots$		$\vdots\vdots\vdots$		$\vdots$	$\vdots$
m	x_{m1}	x_{m2}	$\cdots$	x_{mk}	$\cdots$	x_{mn}	C
column vectors	$\mathbf{x}_1$	$\mathbf{x}_2$	$\cdots$	$\mathbf{x}_k$	$\cdots$	$\mathbf{x}_n$	$\begin{pmatrix} C \\ \vdots \\ C \end{pmatrix} = C\mathbf{1}$
aggregates	$f_1(\mathbf{x}_1)$	$f_2(\mathbf{x}_2)$	$\cdots$	$f_k(\mathbf{x}_k)$	$\cdots$	$f_n(\mathbf{x}_n)$	C

that we may suppose

$$\sum_{k=1}^{n} r_k = \sum_{k=1}^{n} t_k = C \text{ (constant)},$$

while we leave the federation's largesse,

$$\sum_{k=1}^{n} s_k = S,$$

variable. In addition to keeping n constant, we thus generalized previous models in two other respects: The final amount at the k'th state's disposal may be calculated in different ways for different states (previously $f_1 = f_2 = \cdots = f_n$ was supposed) and the total tax income (of all states) can be fixed as C (for another total, one can make a new calculation with the new C). That the *sum* of the hypothetical and the actual tax revenues be the same, can be achieved, if needed, for instance by proportional or additive adjustments. The calculations lead to *Pexider equations for 3-place functions* partly *on essentially restricted domains (triangles)*, partly *restricted only by nonnegativity*. Under two quite natural and weak conditions (that at least one f_j be locally nondecreasing in its first two variables and that $\sum_{k=1}^{n} f_k(0, 0, 0) = 0$), the general solution turns out to be of the form

$$f_k(r, t, s) = (1 - \gamma)r + \gamma t + s + \beta_k \ (k = 1, 2, \ldots, n) \text{ where } 0 \le \gamma \le 1 \text{ and } \sum_{k=1}^{n} \beta_k = 0.$$

Table 2

Data	States 1	2	$\cdots$	k	$\cdots$	n	Sums
real tax revenue	r_1	r_2	$\cdots$	r_k	$\cdots$	r_n	C
hypothetical tax revenue	t_1	t_2	$\cdots$	t_k	$\cdots$	t_n	C
federal grant (planned?)	s_1	s_2	$\cdots$	s_k	$\cdots$	s_n	S
"equalized" funds	$f_1(r_1, t_1, s_1)$	$f_2(r_2, t_2, s_2)$	$\cdots$	$f_k(r_k, t_k, s_k)$	$\cdots$	$f_n(r_n, t_n, s_n)$	$C + S$
equalization payments	$f_1(r_1, t_1, s_1) - r_1$	$f_2(r_2, t_2, s_2) - r_2$	$\cdots$	$f_k(r_k, t_k, s_k) - r_k$	$\cdots$	$f_n(r_n, t_n, s_n) - r_n$	S

Notice the partly weighted structure, the independence of γ from k, and the facts that, for different k, the f_k differ only in constants adding up to 0 ("zero-sum-game"), while the result depends upon C (and n) only through the "constants" (parameters) γ, β_k (for a different C these constants may turn out different, same for a change of n which is more common for municipalities than for states).

The mathematical answer to such questions is not always problem-free. For the original allocation problem the more general equation

$$\sum_{k=1}^{n} f_k(\mathbf{X}) = C,$$

where $f_k: \mathbb{R}_+^{n^2} \to \mathbb{R}_+$ $(k = 1, \ldots, n)$ and all row sums of $\mathbf{X}$ equal C, the general solution is too general for applications. The case where f_k depends only on the $(k-1)$-st, k-th and $(k+1)$-st column of $\mathbf{X}$ (allocation depends upon recommendations not only for that project but also concerning the two neighbouring ones; k is taken modulo n) seems to be unsolved.

For a third application, where (Table 3) x_{jk} is the k-th (kind of) input for the j-th producer $(j = 1, \ldots, m; k = 1, \ldots, n)$ which uses its inputs to produce products of (at most) $y_j = g_j(x_{j1}, \ldots, x_{jn})$ $(g_j: \mathbb{R}_+^n \to \mathbb{R}_+; j = 1, \ldots, m)$ value, the question is the following. Do "aggregator" functions $f_k: \mathbb{R}_+^m \to \mathbb{R}_+$ $(k = 1, \ldots, n)$, $F: \mathbb{R}_+^m \to \mathbb{R}_+$ exist, which collect the k-th kind of input x_{jk} $(j = 1, \ldots, m)$ of all producers as $f_k(x_{1k}, \ldots, x_{mk})$ and the outputs $y_1, \ldots, y_m$ to obtain $F(y_1, \ldots, y_m)$, and a "macroeconomic" function $G: \mathbb{R}_+^n \to \mathbb{R}_+$ collecting the aggregated inputs so

Table 3

Producers	Inputs (goods, services)				Row vectors	Outputs (production functions)
	1	$\cdots$	k	$\cdots$	n	
1	x_{11}	$\cdots$	x_{1k}	$\cdots$	x_{1n}	$\mathbf{x}_1'$	$y_1 = g_1(\mathbf{x}_1')$ $= g_1(x_{11}, \ldots, x_{1n})$
$\vdots$ j	$\vdots$ x_{j1}	$\vdots$ $\cdots$	$\vdots$ x_{jk}	$\vdots$ $\cdots$	$\vdots$ x_{jn}	$\vdots$ $\mathbf{x}_j'$	$y_j = g_j(\mathbf{x}_j')$ $= g_j(x_{j1}, \ldots, x_{jn})$
$\vdots$ m	$\vdots$ x_{m1}	$\vdots$ $\cdots$	$\vdots$ x_{mk}	$\vdots$ $\cdots$	$\vdots$ x_{mn}	$\vdots$ $\mathbf{x}_m'$	$y_m = g_m(\mathbf{x}_m')$ $= g_m(x_{m1}, \ldots, x_{mn})$
column vectors	$\mathbf{x}_1$	$\cdots$	$\mathbf{x}_k$	$\cdots$	$\mathbf{x}_n$		$\mathbf{y}$
aggregates	z_1 $= f_1(\mathbf{x}_1)$	$\cdots$	z_k $= f_k(\mathbf{x}_k)$	$\cdots$	z_n $= f_n(\mathbf{x}_n)$	$\mathbf{z}'$	$F(\mathbf{y}) = F(y_1, \ldots, y_m) =$ $F(g_1(\mathbf{x}_1'), \ldots, g_m(\mathbf{x}_m')) \overset{?}{=}$ $\overset{?}{=} G(f_1(\mathbf{x}_1), \ldots, f_n(\mathbf{x}_n))$ $= G(z_1, \ldots, z_n) = G(\mathbf{z}')$

that the "diagram commutes", that is,

$$G[\,f_1(\mathbf{x}_1), \ldots, f_n(\mathbf{x}_n)] = F[g_1(\mathbf{x}_1'), \ldots, g_m(\mathbf{x}_m')], \tag{1}$$

where $\mathbf{x}_k$ is the k-th column vector, $\mathbf{x}_j'$ is the j-th row vector of the matrix $\mathbf{X} = (x_{jk})$. Clearly this happens exactly if the above functional equation has solutions. If the output values can be added (Aczél–Eichhorn, unpublished, but compare [4], [5]) then

$$F(y_1, \ldots, y_n) = y_1 + \cdots + y_n$$

and the functional equation reduces to

$$G[\,f_1(\mathbf{x}_1), \ldots, f_n(\mathbf{x}_n)] = g_1(\mathbf{x}_1') + \cdots + g_m(\mathbf{x}_m'). \tag{2}$$

If all inputs are "totally separated" then it may be possible to choose also the other aggregator functions as *sums* and we get

$$G(x_{11} + \cdots + x_{m1}, \ldots, x_{1n} + \cdots + x_{mn}) = G(\mathbf{x}_1' + \cdots + \mathbf{x}_m')$$

$$= g_1(\mathbf{x}_1') + \cdots + g_m(\mathbf{x}_m'), \tag{3}$$

a *Pexider equation* on domain restricted only by nonnegativity, with the solution (since $g_1, \ldots, g_m : \mathbb{R}_+^n \to \mathbb{R}_+$)

$$G(\mathbf{x}') = \mathbf{a} \cdot \mathbf{x}' + b_1 + \cdots + b_m, \qquad g_j(\mathbf{x}') = \mathbf{a} \cdot \mathbf{x}' + b_j$$

($\mathbf{a} \geq 0$, $b_j \geq 0$, $j = 1, \ldots, m$). However, *there exist production functions g_j in the economic practice which are not even "close" to affine*. Examples are the popular Cobb–Douglas production functions given by (see e.g. [6])

$$g_j(\mathbf{x}') = g_j(x_1, \ldots x_n) = a \prod_{k=1}^{n} x_k^{c_{jk}}.$$

In general, "the diagram may not commute". Not even relaxing (3) to closeness of the left and right sides helps, because of the *stability of the Pexider equation*. This result of Nikodem [7] holds for monoids and Banach spaces but a consequence for $g_1, g_2, G : \mathbb{R}_+^n \to \mathbb{R}_+$ is that (we take here $m = 2$)

$$\left| G(\mathbf{x} + \mathbf{y}) - g_1(\mathbf{x}) - g_2(\mathbf{y}) \right| \leq \varepsilon \qquad (\mathbf{x}, \mathbf{y} \in \mathbb{R}_+^n) \tag{3'}$$

implies the existence of $\mathbf{a} \geq 0$, $b_1 \geq 0$, $b_2 \geq 0$ such that

$$\left| g_1(\mathbf{x}) - \mathbf{a} \cdot \mathbf{x} - b_1 \right| \leq 3\varepsilon, \qquad \left| g_2(\mathbf{x}) - \mathbf{a} \cdot \mathbf{x} + b_2 \right| \leq 3\varepsilon,$$

$$\left| G(\mathbf{x}) - \mathbf{a} \cdot \mathbf{x} - b_1 - b_2 \right| \leq 4\varepsilon$$

which proves our above point that *only "close to affine" functions can satisfy* (3'). However, in the recent paper of Chmieliński and Tabor [8], there is an example showing that *the inequality*

$$\left| G(\mathbf{x} + \mathbf{y}) - g_1(\mathbf{x}) - g_2(\mathbf{y}) \right| \geq \varepsilon \min\{G(\mathbf{x} + \mathbf{y}), g_1(\mathbf{x}) + g_2(\mathbf{y})\}, \tag{3''}$$

weaker than (3'), *may hold* also *for "very non-affine"* — in fact quite arbitrary bounded — *functions*. Maybe somewhere in-between there is a stability result for this application.

For the more general equation (2) and, even more so, for (1), to given production functions $g_1, \ldots, g_m$, aggregators $f_1, \ldots, f_n$ (and F) and a macroeconomic function G may possibly be found so that the "deficit"

$$\left| G(f_1(\mathbf{x}_1), \ldots, f_n(\mathbf{x}_n)) - F(g_1(\mathbf{x}_1'), \ldots, g_m(\mathbf{x}_m')) \right|$$

be small (with general F or with $F(y, \ldots, y_n) = y_1 + \cdots + y_n$) in the sense of (3′) or (3″). This too is unsolved (but see "Added in proof" at the end of this paper).

Probabilistic models (including stochastic choice models) are becoming popular among mathematical psychologists and people interested in group decision making. Also these lead (or should lead) to functional equations, some to equations quite similar to those which came up above. For others, see [9].

2. Pexider equations, both on "essentially" and on "not essentially" restricted domains, play also prominent roles in utility measurement, uniqueness and comparison problems. An easy example ([10]) is that

$$U_i(x) - U_j(y) = c \Leftrightarrow V_i(x) - V_j(y) = c \qquad (i, j \in \{1, 2, \ldots, n\})$$

$(x, y$ in an arbitrary set Ω; U_k, V_k real valued; $k = 1, \ldots, n)$ has only

$$V_k(x) = U_k(x) + b \qquad (k = 1, \ldots, n)$$

as solutions, while

$$U_i(x) - U_j(x) \geq U_i(z) - U_j(w) \Leftrightarrow V_i(x) - V_i(y) \geq V_j(z) - V_j(w)$$

$$(x, y, z, w \in \Omega; i, j \in \{1, \ldots, n\})$$

$(U_k, V_k : \Omega \to \mathbb{R}; k = 1, \ldots, n)$ has, if card $\Omega \geq 2$, also solutions other than

$$V_k(x) = aU_k(x) + b_k \qquad (k = 1), \ldots, n.$$

(For $\Omega = \{x_1, x_2\}$, for instance, this: $U_1(x_1) = 5$, $V_1(x_1) = 6$, $U_2(x_2) = V_1(x_2) = 0$, $U_\ell(x_1) = V_\ell(x_1) = 1$, $U_\ell(x_2) = V_\ell(x_2) = 0$; $\ell = 2, \ldots, n$.)

Similar implications and equations (and inequalities) come up in the problem of merging scores according to different benchmarks ([11], [12]), not unrelated to aggregation problems (compare section 1) and to measurement problems (implication (8) in section 4).

The above "equivalence formulae" are related to the somewhat older "equality in sacrifice" (for instance for taxation) problems. If $U(x)$ and $U(y)$ are the utilities of the gross income x and of the net income (diminished by taxes) y, respectively, the "sacrifice" is $U(x) - U(y)$. The condition was that, if for gross and net incomes x, z and y, w the sacrifice is the same, then it should also be the same for λ-fold incomes $\lambda x, \lambda y, \lambda z, \lambda w$:

$$U(x) - U(y) = U(z) - U(w) \Leftrightarrow U(\lambda x) - U(\lambda y) = U(\lambda z) - U(\lambda w) \qquad (4)$$

([13], [6]). This used to be motivated by "scale invariance" (a "dimensional" argument of the kind which we intend to question in section **4**). A better argument is the following. Staying with the tax example, if

$$U(x) - U(y) = U(z) - U(w) = s \qquad (5)$$

then the taxes are

$$T(x, s) = x - y, \qquad T(z, s) = z - w \qquad (6)$$

(under supposition of continuity and strict monotonicity of U we can calculate, from (5), y as function of x and s, and w as function of z and s). While linear homogeneity of $T(x, s)$ in x would permit only proportional taxation, the following *generalized homogeneity* seems appropriate. To every multiplier λ and parameter s there exists a parameter $\tilde{s} = S(\lambda, s)$ such that $t = T(x, s)$ implies $\lambda t = T(\lambda x, \tilde{s})$, that is,

$$T(\lambda x, \tilde{s}) = \lambda T(x, s).$$

Then, as in (6),

$$\lambda x - \lambda y = T(\lambda x, \tilde{s}) \quad \text{and} \quad \lambda z - \lambda w = T(\lambda z, \tilde{s}) \qquad (6')$$

and, from (5),

$$U(x) - U(x - T(x, s)) = s.$$

Furthermore,

$$U(\lambda z) - U(\lambda z - T(\lambda z, \tilde{s})) = U(\lambda x) - U(\lambda x - T(\lambda x, \tilde{s})) = \tilde{s}.$$

so that (5) and (6') imply

$$U(\lambda z) - U(\lambda w) = U(\lambda z) - U(\lambda z - T(\lambda z, \tilde{s})) = \tilde{s} = U(\lambda x) - U(\lambda x - T(\lambda x, \tilde{s}))$$
$$= U(\lambda x) - U(\lambda y),$$

that is, (4) indeed holds. The general strictly monotonic (or just nonconstant) continuous solutions of (4) are

$$U(x) = a \log x + b \quad \text{and} \quad U(x) = ax^c + b$$

$(a \neq 0, c \neq 0, b$ arbitrary constants).

3. Speaking of homogeneous functions, a seemingly special question, *when homogeneous functions are increasing*, turned out to be of quite general and deep interest and surprisingly difficult. The analysis of properties of automorphism groups of structures to delineate *broad classes of models that have numerical representation* have lead Cohen, Luce and Narens ([14], [15], [16], [17]) to binary operations with the numerical representation

$$x \circ y = yf(x/y) \quad \text{where } f \text{ is strictly increasing and}$$

$$x \mapsto f(x)/x \text{ strictly decreasing.}$$

So "$\circ$" is clearly a linearly homogeneous strictly increasing two-place function. This leads these authors to the question of *linearly homogeneous strictly increasing multiplace functions* which cannot be so easily characterized, and to the related even harder problem ([15], [18]) concerning the *increasing solutions* of

$$f(r\mathbf{x} + s\mathbf{1}) = rf(\mathbf{x}) + s \quad (r > 0, s \in \mathbb{R}, f\colon \mathbb{R}^n \to \mathbb{R}, \mathbf{1} = (1, \ldots, 1)).$$

By rather involved techniques, these problems have been completely solved in [19].

The same paper determines also necessary and sufficient conditions for the solutions of twelve types of functional equations, describing "*laws of natural and social sciences*" or of "*measurement*", determined in [20], to be increasing. A typical form of such laws is

$$f(\lambda\mathbf{x}) = F(\lambda)f(\mathbf{x}) + G(\lambda)$$

$$(F\colon \mathbb{R}_+ \to \mathbb{R}_+, G\colon \mathbb{R}_+ \to \mathbb{R}, f\colon \mathbb{R}_+^n \to \mathbb{R}, \lambda \in \mathbb{R}_+^n, \mathbf{x} \in \mathbb{R}_+^n). \tag{7}$$

4. Such laws are often explained by "scale arguments" mentioned in connection with (4), like "it does not matter whether lengths are measured in meters or feet", etc. True enough, but there is no guarantee that the *same function* applies to meters as to feet (for instance, the area of a sector of the circle will be measured by a function slightly different from $\alpha \mapsto r^2\alpha$ if α is measured in degrees rather than radians). All one can really say is that, if the dependence of y upon $\mathbf{x}$, measured in a certain unit, is described by the function f, and we change the unit so that the measure of $\mathbf{x}$ becomes $\lambda\mathbf{x}$, then there exists a function g_λ such that

$$y = f(\mathbf{x}) = g_\lambda(\lambda\mathbf{x}).$$

Writing $z = \lambda x$, $v := 1/\lambda$, $H(z, v) := g_{1/v}(z)$, we get only

$$f(vz) = H(z, v),$$

a very weak functional equation. (It says something about H, that it depends only upon the product of its variables, but almost nothing about f). The form of H has to be better specified from outside information, as in the above example (7), where

$$H(z, v) = F(v)f(z) + G(v).$$

An assumption, which avoids the pitfalls of "scale" reasoning or makes it exact and is sufficiently general, could be the following (compare (4)):

$$f(x) = f(z) \Rightarrow f(\lambda x) = f(\lambda z). \tag{8}$$

This can be explained both by "*size*" ("*automorphism of structure*") *and by* "*scale*" ("*dimensional*") *arguments*, the latter being that, while the "independent variable(s)" change according to a "ratio scale", the "dependent variable" has at least a "nominal scale". The implication (8) means that $f(\lambda x)$ depends only upon $f(x)$ and λ:

$$f(\lambda x) = \Phi(f(x), \lambda). \tag{9}$$

By repeated application we get

$$\Phi[f(x), \mu v] = f(\mu v x) = \Phi[f(\mu x), v] = \Phi(\Phi[f(x), \mu], v),$$

that is, Φ satisfies the *multiplicative translation equation*

$$\Phi(\iota, \mu v) = \Phi(\Phi(\iota, \mu), v).$$

Till now we did not specify what kind of quantities x, z, t, λ, μ, v are: A typical situation is where $x, z \in \mathbb{R}^n_+$ (or in a subregion), $f: \mathbb{R}^n_+ \to \mathbb{R}_+$ (accordingly, $t \in f(\mathbb{R}^n_+)$),

$$\mu \in R_{k,m} := \{(\underbrace{\lambda_1, \ldots, \lambda_1}_{k_1}, \ldots, \underbrace{\lambda_m, \ldots, \lambda_m}_{k_m}) \mid \lambda_j \in \mathbb{R}_+; j = 1, \ldots, m\},$$

where

$$k = (k_1, \ldots, k_m) \quad \text{and} \quad k_1 + \cdots + k_m = n,$$

and λ and v are from the same set (we multiply vectors componentwise). What one hopes for are theorems giving a unique solution of the form

$$\Phi(t, \lambda) = \phi^{-1}(\phi(t)\lambda_1^{c_1}\lambda_2^{c_2} \cdots \lambda_m^{c_m})$$

($\phi: f(\mathbb{R}_+^n) \to \mathbb{R}_+$ continuous and strictly monotonic), because then, from (9),

$$\phi \circ f(\lambda x) = \phi \circ f(x)\lambda_1^{c_1}\lambda_z^{c_2} \cdots \lambda_m^{c_m}. \qquad (10)$$

that is, $\phi \circ f$ is an "almost homogeneous function", which is what we expect in "dimensional analysis", though we would have preferred it for f rather than for $\phi \circ f$. Results of Moszner ([21], [22]) show that this holds true essentially if Φ, in addition to being continuous, is *transitive* ($\forall t, u \in f(\mathbb{R}_+^n) \; \exists \lambda: \Phi(t, \lambda) = u$). The case $k_1 = n$ shows that this is a strong condition: in view of (9) it essentially says, in that case, that *f assumes all its possible values already on the ray* $\{\lambda x \mid \lambda \in \mathbb{R}_+\}$. The general continuous solutions have been found also without transitivity ([6], [21]) but these and even the differentiable ones ([23]) are of quite complicated structure (somewhat similar to (11) below). For the other extreme, $k_1 = k_2 = \cdots = k_n = 1$, the transitivity requirement is natural, see again (9): it is satisfied if the equation $\lambda x = y$ has a solution $\lambda = (\lambda_1 \lambda_2, \dots, \lambda_n)$. In this case the problem is readily solved (we have indeed (10)). The solution is more complicated in the remaining cases.

Actually ([24]), one can handle (8) directly without recourse to (9). A typical result is the following.

Let $f: \mathbb{R}_+^n \to \mathbb{R}_+$ *satsify* (8) *for* $x, z \in \mathbb{R}_+^n, \lambda \in R_{k,m}$. *There exists a decomposition* $f(\mathbb{R}_+^n) = \bigcup_{j \in J} E_j$, *bijections* $\phi_j: E_j \to \mathbb{R}_+ (j \in J)$ *and constants* $c_1, c_2, \dots, c_m$ *such that*

$$\phi_j \circ f(\lambda x) = \phi_j \circ f(x)\lambda_1^{c_1}\lambda_2^{c_2} \cdots \lambda_m^{c_m} \qquad (x \in E_j, j \in J, \lambda \in R_{k,m}) \qquad (11)$$

if, and only if, the set

$$E(x) = \{(\ln \lambda_1, \dots, \ln \lambda_m) \mid f(\lambda x) = f(x)\} \qquad (x \in \mathbb{R}_+^n), \qquad (12)$$

is independent of x *and either forms an* $(m - 1)$-*dimensional hyperplane through* 0 *or equals* $\mathbb{R}^m$, *and* $f(\mathbb{R}_+^n)$ *is of the cardinality of the continuum.*

As corollary we have, as desired, the following. *A homeomorphism* $\phi: f(\mathbb{R}_+^n) \to \mathbb{R}_+$ *and constants* $c_1, \dots, c_m$, *not all* 0, *exist so that* (10) *is satisfied iff* $\lambda \mapsto f(\lambda x)(x \in \mathbb{R}_+^n)$ *is continuous,* $f(\mathbb{R}_+^n)$ *is an interval,* (8) *holds, there exists an* $x \in \mathbb{R}_+^n$ *such that* $E(x)$ *(see* (12)*) forms an* $(m - 1)$-*dimensional hyperplane through* 0 *and there exists no* $\mathbf{x} \in \mathbb{R}^n_+$ *such that* $f(\lambda \mathbf{x}) = f(\mathbf{x})$ *for all* $\lambda \in \mathbb{R}_{\mathbf{k},m}$ (which would be the $c_1 = c_2 = \cdots = c_m = 0$ case of (11)). — We have also the following *alternative condition, necessary and sufficient for* (10) *to hold with a homeomorphism* ϕ *and with constants* $c_1, \ldots, c_m$ *not all zero: The nonconstant function f should satisfy* (8), *there should exist an* $\ell \in \{1, \ldots, m\}$ *and an* $\mathbf{x} \in \mathbb{R}^n_+$ *such that* $\lambda \mapsto f(\lambda \mathbf{x})$ *be continuous on* R_ℓ *and on* $\tilde{R}^+_\ell$ *and every value of u on* $\mathbb{R}^n_+$ *be already assumed on* $R_\ell \mathbf{x}$, *where*

$$ R_\ell := \{(\underbrace{1, \ldots, 1}_{k_1}, \ldots, \underbrace{1, \ldots, 1}_{k_{\ell-1}}, \underbrace{t, \ldots, t}_{k_\ell}, \underbrace{1, \ldots, 1}_{k_{\ell+1}}, \ldots, \underbrace{1, \ldots, 1}_{k_m}) \mid t \in \mathbb{R}_+ \} $$

and

$$ \tilde{R}_\ell := \{\lambda \in R_{k,m} \mid \lambda_j = 1 \ (j = n_1 + \ldots + n_{\ell-1} + 1, \ldots, n_1 + n_\ell)\}. $$

There are many more applications and unsolved problems leading to functional equations in the *theory of measurements*.

Added in proof: The equation (1) does have exact (not just approximate) solutions with realistic (e.g. Cobb–Douglas) production functions but not when F (or any of $f_1, \ldots, f_n$) is the addition. See a forthcoming paper of J. Aczél and G. Maksa, cf. also [5] and [4].

Acknowledgement

This research has been supported in part by Natural Sciences and Engineering Research Council of Canada grants.

REFERENCES

[1] ACZÉL, J., NG, C. T., and WAGNER C., *Aggregation theorems for allocation problems.* SIAM J. Algebraic Discrete Methods 5 (1984), 1–8.
[2] BUHL, U. H. and PFINGSTEN, A., *Eigenschaften und Verfahren für einenangemessenen Länder-finanzausgleich in der Bundersrepublik Deutschland.* Finanzarchiv 44 (1) (1986), 98–109.
[3] ACZÉL, J. and PFINGSTEN, A., *Constituent-sensitive public fund sharing.* In Mathematical modelling in economics. Springer, Berlin, 1993, pp. 3–10.
[4] POKROPP, F., *Aggregation von Produktionsfunktionen. Klein-Nataf-Aggregation ohne Annahmen über die Differenzierbarkeit und Stetigkeit.* [Lecture Notes in Econom. and Math. Systems, No. 74]. Springer, Berlin, 1972.
[5] VAN DAAL, J. and MERKIES, A. H. Q. M., *Consistency and representativity in a historical perspective.* In *Measurement in economics.* Physica, Heidelberg, 1987, pp. 607–637.

[6] ACZÉL, J., *A short course on functional equations based upon recent applications to the social and behavioral sciences*. Reidel (Kluwer), Dordrecht-Boston, 1987.

[7] NIKODEM, K., *The stability of the Pexider equation*. Ann. Math. Sil. *5* (1991), 91–93.

[8] CHMIELIŃSKI, J. and TABOR, J., *On approximate solutions of the Pexider equations*. Aequationes Math. *46* (1993).

[9] MARLEY, A. A. J., *A selective review of recent characterizations of stochastic choice models using distribution and functional equation techniques*. Math. Social Sci. *23* (1992), 5–29.

[10] BOSSERT, W., *On intra- and interpersonal utility comparisons*. Soc. Choice Welf. *8* (1991), 207–219.

[11] ROBERTS, F. S., *Merging relative scores*. J. Math. Anal. Appl. *147* (1990), 30–52.

[12] ACZÉL, J., *Determining merged relative scores*. J. Math. Anal. Appl. *150* (1990), 20–40.

[13] YOUNG, H. P., *Progressive taxation and the equal sacrifice principle*. J. Public Econom. *32* (1987), 203–214.

[14] COHEN, M. and NARENS, L., *Fundamental unit structures: a theory of ratio scalability*. J. Math. Psych. *20* (1979), 193–232.

[15] LUCE, R. D., *Rank dependent subjective expected-utility representations*. J. Risk Uncert. *1* (1988), 305–332.

[16] LUCE, R. D. and NARENS, L., *Classification of concatenation structures by scale types*. J. Math. Psych. *29* (1985), 1–72.

[17] NARENS, L., *A general theory of ratio scalability with remarks about the measurement-theoretic concept of meaningfulness*. Theory and Decision *13* (1987), 1–70.

[18] ACZÉL, J., *Three problems*. In *General inequalities 6*. Birkhäuser, Basel-Boston-Berlin, 1992, p. 477.

[19] ACZÉL, J., GRONAU, D., and SCHWAIGER, J., *Increasing solutions of the homogeneity equation and of similar equations*. J. Math. Anal. Appl. *182* (1994), 436–464.

[20] ACZÉL, J., ROBERTS, F. S., and ROSENBAUM, Z., *On scientific laws without dimensional constants*. J. Math. Anal. Appl. *119* (1986), 389–416.

[21] MOSZNER, Z., *Une généralisation d'un résultat de J. Aczél et M. Hosszú sur l'équation de translation*. Aequationes Math. *37* (1989), 267–278.

[22] MOSZNER, Z., *Sur une forme de la solution de l'équation de translation*. C.R. Math. Rep. Acad. Sci. Canada *12* (1991), 285–290.

[23] FÖRG-ROB, W., *Differentiable solutions of the translation equation*. [Grazer Math. Ber., No. 287]. Karl-Franzens Univ., Graz, 1988.

[24] ACZÉL, J. and MOSZNER, Z., *New results on "scale" and "size" arguments justifying invariance properties of empirical indices and laws*. Math. Social Sci. *28* (1994), 3–33.

Department of Pure Mathematics,
University of Waterloo,
Waterloo, Ont. N2L 3G1,
Canada.

Aequationes Mathematicae **50** (1995) 50–72
University of Waterloo

0001–9054/95/020050–23 $1.50 + 0.20/0

Progress of iteration theory since 1981

GYÖRGY TARGONSKI

Introduction

This survey tries to highlight a number of recent developments in iteration theory, and to point out a number of unsolved problems, thus also trying to predict the direction the evolution may take.

At least two things in this approach are arbitrary. "Recent" was chosen to mean "since about 1981", when [Targonski 81] was published, to the best of my knowledge the first book on iteration theory in general.

The choice of the topics is also, by necessity, somewhat arbitrary. Obviously I am talking more about the fields I know more about. In some cases, there are obvious objective reasons for the choice. For instance, numerical methods are in large part based on iteration; this immense field no longer can be counted as part of iteration theory proper. Also, one-dimensional discrete dynamics is now a field in its own right; while I devoted a part of [Targonski 81] to it, now I will only give references. I can however promise the following. The "iterated list of references", that is, the union of the lists of references in the books and papers listed in this paper united with the list of references itself, does contain a large part of what has been done in iteration theory in recent years.

The following fields within iteration theory will be treated.

(1) *Orbit theoretical iteration* theory, that is, study of the structure imprinted upon a set by a given self-mapping.
(2) *Algebraic iteration theory*.

AMS (1991) subject classification: Primary 39B12, Secondary 26A18, 58F08.

Manuscript received November 12, 1992 and, in final form, October 13, 1993.

(3) *Iteration of formal power series.*

(4) *The Liedl transformation* (Pilgerschritt transformation) as a method of embedding a function in a one-parameter group of functions where the parameter takes on all real values (continuous iteration group).

(5) *"Aczél–Jabotinsky dynamics".* The three Aczél–Jabotinsky equations can be derived from the translation equation but not, in general, vice versa. Thus "weak dynamical systems" arise which have some but not all properties of a true dynamical system satisfying the translation equation.

(6) *The functional equations of Abel and of Schröder. Commuting functions. Real iterates.*

(7) *Cellular automata.* These automata, discovered by von Neumann and Ulam and then half forgotten now have a renaissance, with many applications. Their iteration (cellular automata are discrete, autonomous semi-dynamical systems!) poses many problems; there are interesting results.

(8) *Functional analysis.* Iteration of a function can be discussed by (crudely speaking) considering the linear operator of right composition with the function. This leads to new insights and results.

(9) *Phantom iterates.* Since functions in general have no iterative roots (fractional iterates) of every order (and thus no continuous time iterate) "generalized embeddings" have been sought for a long time. Since 1984, the idea of phantom iterates offers one such approach.

(10) An *Appendix* briefly discusses various topics which are outside the main fields outlined above but should be mentioned in our survey.

1. Orbit theory

In orbit theory we look solely at the structure imprinted on a set S by a self-mapping f. There is no other (algebraic, topological, . . .) structure on S, which now is the union of (Kuratowski–Whyburn) orbits of f; for the simple properties of orbits see e.g. [Targonski 81], [Targonski 84], chapters 1 and 2.

We can give a few examples of purely "orbit theoretical" results. We need the notion of ultrastability ([Sklar 1969]). f is called ultrastable if $f|f(s)$ is bijective. Ultrastability is "almost as good as bijectivity", as the following result shows ([Weitkämper 85]): a mapping can be embedded in a $\mathbb{Z}$-group if and only if it is ultrastable.

Curiously, the perhaps best known unsolved problem in iteration theory appears in a purely orbit theoretical context. It is the $3x + 1$-problem, also called the Collatz problem, or the Ulam problem, or the Syracuse problem, or the Kakutani problem; Hasse's algorithm is also a usual term. We state the conjecture in the following simple form, following [Wagon 85]. Given the self-mapping f of $\mathbb{N}$ onto itself

$$f(n) = \begin{cases} \dfrac{n}{2} & (n \text{ even}) \\ 3n + 1 & (n \text{ odd}) \end{cases}$$

one sees at once that $(1, 4, 2)$ is a 3-cycle of f.

Conjecture: every splinter (iteration sequence) of f terminates in the cycle (1, 4, 2). Much numerical work has gone into the problem, stochastical methods were used, the question of decidability raised. For a survey see [Lagarias 85]. Interestingly, the problem can be reformulated so as to involve chaos and fractals ([Agnes, Rasetti 88]). The problem was unsolved in 1991 ([Gale 91]).

2. Algebraic iteration theory

It seems clear, that many problems in "orbit theory" (cf. Section 1) could be treated by algebraic methods. A good example is the case of iterative square roots of self-mappings of arbitrary sets; the solution was given by [Isaacs 50]. Since the set of all self-mappings of a set is a semigroup, finding square roots of elements in a semigroup is a generalization of the Isaacs problem. This problem was solved in [Snowden, Howie 82] for the case of finite sets. It would be interesting to see whether the general solution of the iterative root problem in [Riggert 75], see also [Targonski 81], Section 2.1, (roots of arbitrary order on an arbitrary set) could be treated in the Snowden–Howie style.

An approach to "algebraization" of iteration could possibly be through unary algebras with research results already in the nineteen-sixties. As starting point one could take [Skornjakov 77] (with 41 references!) as well as [Chvalina, Matoušková 84] and [Blažková, Chvalina 84].

Since orbits are equivalence classes of an equivalence relation (existence of a common successor), recent work [Schleiermacher 93] in the direction of the Krasner theorem on invariant relations ([Krasner 38]) could become a tool in orbit theory.

Following the pioneering work of Carlo Bourlet ([Bourlet 97_1, 97_2]) right composition operators $T\varphi := \varphi \circ f$ (defined on suitable function spaces) have become part of iteration theory. It is of interest also to consider (nonlinear) left composition operators ($A\varphi := \alpha \circ \varphi$) and so on. An attempt to systematize all this is in [Targonski 90]. Results on algebraic right composition operators are in [Böttcher, Heidler 92].

Last, but not least, linear mappings $\mathbf{r} \mapsto A\mathbf{r}$ ($\mathbf{r} \in \mathbb{R}^n$), where A is an $n \times n$ matrix, pose nontrivial iteration problems for $n \geq 1$. There is of course a large amount of work on this, some by people who did not know they were doing iteration theory (seen from our point of view). A recent contribution (on square roots of uppertriangular matrices) is [Miller 91].

The orbit structure of any self-mapping of a set is invariant under conjugacy. For work in the important field of conjugacy see e.g. [Schweizer, Sklar 88].

3. Iteration of formal power series

Research by L. Reich and his colleagues and collaborators in Graz continued vigorously during the decade we are surveying.

[Reich, Schwaiger 80] introduced a linearization method for the solution of certain functional equations, which later turned out to be important also in another context, interpreted as a phantom iterate (see Section 9). In [Reich 85] the third Aczél–Jabotinsky equation for formal power series in one variable is solved. The relations between the three Aczél–Jabotinsky equations and their relations with the translation equation (5.1) was clarified around that time (see Section 5) in [Reich 88, 89, 91], [Aczél, Gronau 88, 88_1], [Gronau 91, 91_1].

The relationship between families of commuting functions and iteration groups have been investigated also in the context of formal power series ([Reich 88, 89]) see Section 6.

As these examples show, formal power series occupy an important position in iteration theory—many general problems are also treated in this context.

Results of the theory up to the end of 1980 have been surveyed in [Targonski 81], Section 6.2. A description of early results can be found in [Peschl, Reich 71]. An outline of the iteration theoretical aspects of formal power series (more specifically of formal biholomorphic mappings) and some open problems are given below. We closely follow L. Reich (personal communication).

Let $\mathbb{C}[[x]]$ be the ring of formal power series in the indeterminate $x = {}^t(x_1, \ldots, x_n)$ over $\mathbb{C}$, Γ the group of automorphisms F of $\mathbb{C}[[x]]$ continuous in the order topology, moreover $F|_{\mathbb{C}} = id$. The following are the central problems leading to further development.

I. Does there exist, for a given $F \in \Gamma$, a family $(F_t)_{t \in \mathbb{C}}$ in Γ ("iteration of F") such that $F_1 = F$ and $F_t \circ F_s = F_{t+s}$ for every $t, s \in \mathbb{C}$?

II. Does there exist, for a given $F \in \Gamma$ and for a given $r \in \mathbb{N}$ a $G \in \Gamma$ ("r-th iterative root of F") such that $g^r = F$ where g^r denotes the r-th iterate of g?

(1) Building on results of S. Sternberg, N. Lewis and E. Peschl, L. Reich looked for criteria ensuring the existence of analytic iterations for given $F \in \Gamma$ and given choice Λ of logarithms of eigenvalues of the linear part of F. $(F_t)_{t \in \mathbb{C}}$ is called analytic if the coefficients g_r in $r_t(x) = \sum_{r \in \mathbb{N}_0^n} g_r(t) x^r$ are entire functions. The problem was solved by L. Reich and J. Schwaiger using the normal forms under conjugation in Γ in particular the smooth normal forms for 1. The proofs also furnished the construction of the iteration groups and pointed to connections with autonomous differential systems and differential equations with complex linearization, following P. Erdös and E. Jabotinsky.

(2) Analogous results (using the normal forms mentioned above) were found for the existence of iterative roots.

(3) From these criteria grew results on connection between existence of analytic iterations and of "sufficiently many" iterative roots of an $F \in \Gamma$ (results of L. Reich and A. R. Kräuter; cf. (4) below).

(4) If one demands only continuity of the coefficients G_r ("continuous iteration") then the method leads to the normal forms. F is analytically iterable if and only if it is continuously iterable. The continuous iterations of an $F \in \Gamma$ are precisely the real-analytic iterations. (Results of L. Reich and W. Bucher.)

(5) The next question is this. What about iteration with no conditions whatsoever on the coefficients ("iterable F"). G. Mehring and C. Praagman showed, independently, that F is analytically iterable if and only if it is iterable. Using methods of algebraic geometry, C. Praagman even showed the following. If $F \in \Gamma$ has iterative roots of all orders then F is analytically iterable.

(6) In the case of one variable ($n = 1$) the theory of iterations, as sketched above, and results on the set of solutions of the equation

$$(G \circ \Phi)(x) = \frac{d\Phi}{dx} G(x)$$

(the Julia equation, a special case of the third Aczél–Jabotinsky equation) play a decisive part in the explicit description of the families of commuting automorphisms. For $n \geq 2$ this question is open.

(7) Another interesting problem seems to be the distribution of iterable (and of not iterable) automorphisms in the neighbourhood of a given $F \in \Gamma$. "Neighbourhood" may be defined by the order topology or by the coefficient-wise topology. First results in this direction (for $F = id$) were due to S. Sternberg.

So far the concise description of history and unsolved problems suggested by L. Reich. Let me add that [Reich 92, 93₁] are contributions concerning problem (7), about the distribution of iterable functions.

Closing this section let me express my personal feeling that the linearization method [Reich 71], [Reich, Schwaiger 80] will play a continued important part in Phantom Dynamics (cf. Section 9).

4. Liedl's Pilgerschritt transformation

Embedding of functions in "time continuous" iteration (semi-) groups is one of the central themes of iteration theory; it appears in several places also in the present survey.

There is an attempt to solve this problem which has been in the folklore of the theory for many years. It can be explained in the language of the (Bourlet) substitution operators. Assume f, the function to be embedded is a continuous self-map of I, the closed unit interval. Consider the following bounded linear operator A on $C(I)$: $A\varphi = \varphi \circ f$. One may say "Now determine the linear operator $\log A$, then you have the embedding of A: $A^t = e^{t \log A}$. Applying this semigroup of operators to $x \in C(I)$, we find $A^t x = f^t(x)$ and the embedding has been achieved".

This sounds too good to be true and in fact it is not true. The scheme does not work except in special cases. The first problem of course is with the existence of the logarithm of a substitution operator. Trying to sweep the problem under the carpet, we encounter divergence problems. Thus a method not using logarithms would be of advantage. Even so, any embedding would have to have a built-in failure mechanism for the case that the embedding does not exist and in general it does not exist.

R. Liedl's Pilgerschritt transformation is a "logarithm-free" method. Introduced in the nineteen-seventies, it is much more than a method of embedding. It has become a field of research in its own right, branching out in various directions. Work published before the end of 1980 was included in [Targonski 81], with the appropriate references. Since 1980 much new work was done, new concepts emerged and new results were found. We attempt an overview of some of these.

The Liedl transformation (Pilgerschritt transformation) is a method for finding one-parameter subgroups in topological groups. Given an element of the group in the same connected component as the unit element, an arbitrary path between the two is taken, and repeatedly subjected to the Liedl transformation. The resulting sequence of paths may converge to—or even reach in finitely many steps—a "homomorphic path", that is, a one-parameter subgroup.

One way of getting acquainted with Liedl's idea and early work by him and his co-workers, in English, is reading Chapter 4 in [Targonski 81]. The papers [Liedl, Netzer, Reitberger 81, 82] are introductions (in German) to the theory and its results at that stage.

The Pilgerschritt transform of a path $A(t)$ $(0 \le t \le 1)$ is given among others in $GL\ (n, \mathbb{R})$, for the case of a smooth path by

$$\tilde{A}(t) = M(1), \tag{4.1}$$

where M is the unique solution of the initial value problem

$$M'(\tau) = tA'(\tau)A^{-1}(\tau)M(\tau) \qquad M(0) = I. \tag{4.2}$$

In full generality the definition of the transformation is

$$\tilde{A}(t) + \lim_{|\pi| \to 0} \Pi(A; \pi, t) \tag{4.3}$$

where the Liedl product (Pilgerschritt product) Π is defined as

$$\Pi(A; \pi, t) := [A(a^{*}_{m-1})A(a_{m-1})^{-1} \cdots A(a^{*}_{0})A(a_{0})^{-1}]. \tag{4.4}$$

Here π is a partition $0 = a_0 < a_1 < \cdots < a_m = 1$ of the unit interval, $|\pi| = \max a_{k+1} - a_k$ and $a^{*}_k := a_k + t(a_{k+1} - a_k)$ and $t \in [0, 1]$ is fixed. While [Netzer 82] deals with the convergence of the sequence of iterated Pilgerschritt transforms, [Netzer, Reitberger 82] investigates this problem for nilpotent Lie groups. In [Förg-Rob 85] it is shown that the transform can be computed within the Lie-algebra of the Lie group in question. Another part of the paper deals with the Pilgerschritt transform in the group of vectors of formal power series.

In [Förg-Rob, Netzer 85] the following problem is treated. Paths which are already homomorphic are invariant under the Pilgerschritt transformation, that is, fixed points. The decisive (and highly nontrivial) question is whether a given fixed point is attractive or not. This problem is treated in subgroups of the group of invertible matrices, using the method of product integration.

[Liedl 86] treats group valued power series.

[Förg-Rob 89] adds new results for the complex affine group.

[Liedl, Netzer 89] actually contains the material of two lectures at ECIT 87. Liedl's "short ruler" method uses an idea from differential geometry to solve the translation equation (with time-one condition), that is, achieve embedding. Netzer uses product Taylor expansion (PTE), that is, "Taylor products" to achieve the same goal. (For PTE and related topics see also [Cap 89]).

[Cap 91], somewhat different in style and approach compared to the rest of this Section, considers the Abel equation, an equation the author refers to as the (third) (Aczél–) Jabotinsky equation (actually it is a special case, the Julia equation) and the "inverse problem of ordinary differential equations". This last problem is the following. Given a time-one map of a time-continuous flow (it maps every point to the point where it will be in one time unit.) The task is to reconstruct the flow (or only its vector field). Thus we again have the embedding problem. This paper is an interesting meeting point of several topics discussed in the present survey.

[Netzer, Liedl 91] introduces an "improved version" of Liedl's transform: the fast Pilgerschritt transformation (FTP). The starting point is replacing the initial value problem (4.2) by a "better one". [Liedl, Netzer 91] is a more detailed paper of 54 pages, also discussing FTP. See also [Netzer 92], where a one-step Liedl transformation is described: under certain conditions the homomorphic path is reached in one step.

To conclude this section we note that the Liedl transformations may be used in the construction of phantom iterates (see Section 9).

5. The Aczél–Jabotinsky equations. "Weak dynamics"

The translation equation

$$F[F(x, s), t] = F(x, s + t) \tag{5.1}$$

describes an autonomous semi-dynamical system if x is interpreted as a point in the state space X ("phase space") of some "system", and the second variable of F ranges over an additive semigroup of $\mathbb{R}$. Customarily but by no means necessarily

$$F(x, 0) = x \tag{5.2}$$

is stipulated.

(For the general theory of (5.1) and a far-reaching generalization, the transformation equation, see the survey paper by Z. Moszner in this issue.)

The mapping

$$F(x, 1) =: f(x) \tag{5.3}$$

is the "time-one map" of the semidynamical system.

If t ranges over $\mathbb{R}$, then F is bijective for every fixed t and we have a dynamical system. (Often a semidynamical system is called, somewhat sloppily, a dynamical system.)

For example an iteration group $f'(x)$ satisfies (5.1) if we put $F(x, t) = f'(x)$.

The translation equation (5.1) requires no structure on X; a fairly general but still very useful condition is that X be a Banach space. In our present context we take x to be either a real number or a complex number. We present the idea formulated for this case.

As shown in [Aczél 49 and elsewhere] (see also [Jabotinsky 55, 63]), three equations can be derived from the Translation Equation under suitable differentiability conditions. Introducing

$$g(x) := \left. \frac{\partial F(x, t)}{\partial t} \right|_{t=0} \tag{5.4}$$

we find the three Aczél–Jabotinsky equations

$$\frac{\partial F}{\partial t} = g \frac{\partial F}{\partial x} \tag{5.5a}$$

$$\frac{\partial F}{\partial t} = g \circ F \tag{5.5b}$$

and consequently

$$g \frac{\partial F}{\partial x} = g \circ F. \tag{5.5c}$$

An avalanche of research was started by D. Gronau asking the following question (see [Targonski 84], Gronau's problem (3.3.11)): Can the translation equation be deduced from the first or third Aczél–Jabotinsky equation? These questions and related ones were answered in the negative in [Aczél, Gronau 88_1, 88_2]; see also [Gronau 88, 91, 91_2] and general solutions were given of (5.5a,b,c) individually, by pairs and collectively. An attempt of a dynamical interpretation was made in [Targonski 91] (see also [Aczél 91]). Results in this direction are given in [Gronau 91]. It may be of interest to find applications of what could be called "Aczél–Jabotinsky dynamics": time evolutions satisfying the first, (5.5a) or the third, (5.5c) Aczél–Jabotinsky equations or both, with or without one or both conditions (5.2), (5.3).

The second Aczél–Jabotinsky equation (5.5b) is equivalent to the translation equation (5.1) if the initial value problem "(5.1b) with (5.2)" is well posed.

Time evolutions of this kind could emerge for example in biology, or in economics, but perhaps even in physics.

For formal power series the problem was solved in [Reich 91]; see also [Reich 85, 88, 89].

6. Iteration sequences. Groups and semigroups of iterates

Throughout this survey we emphasize new directions of research as well as generalizations. This section, however, is about "hard core, classical" iteration. Typically real functions of one variable are treated, but functions in $\mathbb{R}^n$, even in topological spaces make their appearance. Of course, new ideas and techniques emerge, but the flavour of research is the traditional one.

The first thing to read, of course, is chapter 1, Iteration, of the important book [Kuczma, Choczewski, Ger 90]. Also relevant to our topic in this section are the results about the Schröder and Abel equation, to be found in various sections of the book.

We note that there are two families of functions easy to embed in a group.

For $f(x) = \lambda x$ $(0 < \lambda < 1)$ we have $f^t(x) = \lambda^t x$, and for $g(x) = x + c$ $(c \neq 0)$ we have $g^t(x) = x + ct$. For functions conjugate to a linear transformation $f(x) = \varphi^{-1}[\lambda \varphi(x)]$ or to a translation $g(x) = \psi^{-1}[\psi(x) + c]$ we find the embedding $f^t(x) = \varphi^{-1}[\lambda^t \varphi(x)]$ ("Schröder form") or $g^t(x) = \psi^{-1}[\psi(x) + ct]$ ("Abel form"). Then existence of bijective solutions

$$\varphi[f(x)] = \lambda \varphi(x) \qquad \text{(Schröder equation)} \tag{6.1}$$

or

$$\psi[g(x)] = \psi(x) + c \qquad \text{(Abel equation)} \tag{6.2}$$

in a suitable domain implies embeddability. Thus the Schröder equation and the Abel equation (as well as the seldom occurring Böttcher equation $\varrho[h(x)] = \varrho(x)^a$) are closely linked to the problem of embedding ("continuous iteration" in an older and ambiguous terminology).

We also note that different elements of the same iteration group commute: $f^s \circ f^t = f^{s+t} = f^{t+s} = f^t \circ f^s$. The following question arises. Under what conditions are two commuting bijections "iterates of each other", that is, elements of the same iteration group? The question is even more profitably posed for maximal sets of commuting functions. The study of such families is interesting for its own sake.

All this has been known for a long time. In this section we also quote some new results in this general direction. We cite a few recent papers.

In [Zdun 85], homeomorphisms of the circle are embedded in a time-continuous flow, that is, in a real-parameter group of homeomorphisms.

In [Zdun 89] older results ([Zdun 79], [Smajdor 85]) are generalized: if f^t is an iteration semigroup on a compact metric space and $t \mapsto f^t(x)$ is measurable then it is continuous. This is the main result.

In [Zdun 90], quasi-continuous iteration groups and semigroups of real functions are characterized. These are semigroups $f^t(x)$ such that $f^t(x)$ is a continuous function of t, while no such condition is imposed with respect to x.

[Zdun 91] gives the structure of iteration groups of continuous functions such that the group elements with the exception of the identity mapping have no fixed points. In the representation of the group the "Schröder form" and a generalization of the "Abel form" appear.

In [Zdun 85_1] the representation for regular iteration semigroups $f^t(x) = \lim_{n \to \infty} f^{-n}[\lambda^t f^n(x)]$ appears. It may be interesting to look at this formula under conditions as general as possible. There is a strange formal similarity to the scattering operator in quantum theory.

C^r iteration groups are discussed in [Zdun 89_1].

In [Zdun 91_1], continuous iteration groups of fixed point free mappings in $\mathbb{R}^n$ are treated. The principal result is that these groups can be represented in the Abel form.

[Zdun 88] treats a case where, under appropriate conditions, two commuting functions are—as defined earlier in this Section—"iterates of each other". [Zdun 89_2] deals with simultaneous Abel equations, again leading to commuting functions and representation in the "Abel form" of an iterative group appearing in this context.

For systems of Abel equations and related topics see also [Neuman 82, 89].

[Zdun 92] discusses continuous, strictly increasing, commuting self-mappings of an open set. Relations between the iteration sequences of the two functions are established.

[Smajdor 89, 92] are recent additions to work of the author on the iteration of set-valued functions.

For the characterization of Zdun flows see [Sklar 87], and for the non-embedability of the baker's transformation [Schweizer, Sklar 90].

7. Cellular Automata

Cellular automata were introduced by J. von Neumann and S. Ulam; see the volume [von Neumann 66]. The matter lay dormant for a long time (the posthumous volume cited contains ideas of von Neumann from various times). During the past twenty-odd years, furious research activity sprang up, and now cellular automata is a large and rapidly growing field. For an overview of the field at one stage see [Farmer et al. 85].

A cellular automaton is a countable set (the "grid"; think for instance of a rectangular grid in the plane); each grid point is occupied by one of finitely many symbols (think for instance of 0, 1). Every grid point has a finite neighbourhood, consisting of grid points not necessarily neighbours in the ordinary geometrical sense; the neighbourhood may include the point itself. At every time impulse, every grid point gets a new symbol (possibly the old one) as a function of the "local configuration", that is, the way the neighbourhood is occupied by symbols; this is the local transition function.

A useful property is that every grid point should have the "same" neighbourhood and the same local transition function. These properties are expressed as "shift invariance". If we try to survey all possible or all useful "grids", questions of algebraic topology arise.

The local transition function is not everything, of course. Since all grid points get a new symbol at the "time impulse", the global configuration (a map of the countable set of grid points to the finite set of symbols) also changes. The map from the set of all configurations (the configuration space) to itself is the global transition function.

The problem arises of characterizing those global transition functions which are induced by a local transition function. In [Ferber 91] the following is shown. Introducing an appropriate metric on the configuration space, a self-mapping of the configuration space is the global transition function of a cellular automaton if and only if it is continuous.

It is a striking feature of cellular automata that a very simple local transition rule may give rise to very complicated and interesting global behaviour. The perhaps most famous example is John Horton Conway's "game of life" automaton (see [Gardner 70]).

Why is all this relevant to iteration theory?

From the definition it is clear that a cellular automaton is an autonomous discrete-time semidynamical system. The behaviour of such a system is given by the iteration of a continuous self-mapping of a configuration space. All results of iteration theory—starting with the orbit theoretical results—apply to the time evolution of cellular automata. We give a few examples: fractional iterates of cellular automata can be discussed; see [Ferber et al. 91]. The results on limit sets of continuous mappings [Graw 82, 84] can be applied to the time evolution of cellular automata (see [Langenberg 92]). Orbit theoretical results may be also applied ([Ferber 88]). A suitably modified version of the notion of orbit entropy ([Burkart 82]) can be used to study the behaviour of cellular automata (Langenberg 92]).

The number and scope of applications in computing, gas dynamics (to name only two fields) and so on, is breathtaking. Iteration theory has insights, results and methods which are (unfortunately) still largely unknown to outsiders. The challenge to iteration theory is clear.

8. Functional analysis

In [Bourlet 97, 97_1] the linear operator $F\varphi := \varphi \circ f$ was introduced. Bourlet's idea turned out to be a most fruitful one. The ramifications of this approach have been discussed in [Targonski 67, 81] and are outside the scope of this survey. However, work on the generators and co-generators of substitution semigroups continued until 1987; see [Targonski, Zdun 85, 87].

It seems that at the present time—and for some time to come—the most useful application of functional analysis to iteration theory will be the method of generalized embeddings (in particular, generalized iterative roots) we call phantom iterates (phantom dynamical systems). This will be discussed in Section 9.

9. Phantom iterates

As is well known, the functional equation $g^r = f$ (f, g are self-mappings of a set) has in general no solution;—in other words, the iterative root (fractional iterate) $f^{1/r}$ does not exist. To give a simple example: if f has exactly one 2-cycle ($x \neq y$, $f(x) = y$, $f(y) = x$), then f has no iterative square root. Then there exists no embedding f^t for $f: f^s \circ f^t = f^{s+t}$, $f^1 = f$, $t \geq 0$, since the choice $t = \frac{1}{2}$ would yield a (non-existing) square root. Thus the idea of generalized embedding arose quite naturally. Saying it quite simply: the trouble is that there are too many or too few orbits of a given type. (In our above example, extending the domain by two points forming a 2-cycle would help, unless there are additional obstacles.) In a similar sense, removing the offending 2-cycle would also help; but, in the presence of a topology, there now would be two holes in the domain. So, extension of domain is the better solution, and this idea was in fact carried out; see [Peschl, Reich 71], [Reich, Schwaiger 80], [Mira, Müllenbach 83].

Our approach is different. Formulated for the "maximal" problem, the problem of embedding, it can be crudely formulated as follows. Even if no embedding f^t exists, the Bourlet substitution operator $A\varphi := \varphi \circ f$ may be embeddable in a one-parameter semigroup of operators A^t; then A^t serves as a (weaker) version of the non-existing f^t. This is the phantom iterate, described intuitively and imprecisely. The idea was hinted at but not followed through in [Targonski 81], and introduced (under the temporary name of "weak iterate") in [Targonski 84]. Here the setting is quite general, but only the problem of iterative roots is addressed. The semigroup of all self-mappings of a set is isomorphically immersed in a larger semigroup, so that the equation $g^r = f$ has a solution in the larger semigroup. This approach was demonstrated on the case of self-mappings of a finite set, represented by "mapping matrices" with 0, 1 entries. If the r-th root of such a mapping is not itself a mapping matrix, it is still (in later terminology) a phantom root. This approach was followed up and elaborated in [Bartels 91].

Phantom iterates were formally introduced, rigorously defined and investigated in [Targonski 84₁]. The notion is introduced in a general form and then applied to certain continuous self-maps of the closed unit interval. A class of such functions was given which have non-trivial phantom square roots. This result was generalized in [Krause 88], [Bartels 91] and [Targonski 93].

Phantom iterates have been also established for formal power series. As already mentioned, a construction in [Reich, Schwaiger 80] can be used as phantom iterate, once one has the notion (Reich–Schwaiger phantom). For a thorough discussion, see [Schwaiger 89]; see also [Schwaiger 91].

Phantom roots of cellular automata were treated in [Ferber et al. 91], sections 2.3, 2.4, 4.1, 4.2, 4.3.

A direction in which so far nothing has been done is phantom roots of mappings from $\mathbb{N}$ to itself. As we saw, the finite case has been extensively dealt with, and there are also results on continuous self-mappings of [0, 1]; but $\mathbb{N} \to \mathbb{N}$ is terra incognita.

We conclude this section by describing a fairly general case of phantom iterates in the language of dynamics.

Consider the discrete autonomous semidynamical system (X, f). Here X is a compact topological space, the state space of some system, while f, the "next state function", is a continuous self-mapping of X. We are looking for a phantom embedding of f. Consider now Φ, the family of all real-valued continuous functions on X. (We could take complex valued functions on X, in the general case the functions could take values in any Banach algebra. Multiplication is defined pointwise). We interpret every φ as some special kind of measurement on X, so that $\varphi(x)$ is some real number partly characterizing the particular state $x \in X$ of the system. Introduce now the linear operator $A\varphi := \varphi \circ f$ on Φ, considered as a Banach space, and assume that an embedding of A exists in a semigroup A^t, where $t \geq 0$ is interpreted as time. Consider now the case that, for a particular t (for instance $t = \frac{1}{2}$), f^t does not exist, thus the state $f^t(x)$ does not exist. On the other hand, $(A^t\varphi)(x)$ is a numerical measurement of the non-existing state $f^t(x)$. For all $\varphi \in \Phi$ such a measurement can be carrried out and thus the non-existing state gains a kind of "phantom existence", hence the name.

Staying with the case $t = \frac{1}{2}$, it has been shown in [Targonski 84_1] that for certain f and some restriction on Φ, the fantom is the sum of two substitutions: $(A^{1/2}\varphi)(x) = \varphi[\alpha(x)] + \varphi[\beta(x)]$.

In this case (and in its generalizations) the system behaves as if it were in several states at the same time, and the individual measurement values have to be added to obtain the measurement values on the phantom states. An analogy with quantum mechanics is apparent; physicists have been quick to acknowledge this.

As already hinted at in Section 3, Liedl's transformation can be used to find a phantom embedding for invertible f ([Targonski 94]).

Appendix

In this Appendix, we briefly refer to fields which belong to iteration theory, but for reasons historical and/or practical, could not be included in this survey of research.

I. *Numerical methods.* A large part of this field is based on iteration. As an important example we cite [Deslauriers, Dubuc 91]. This field is immense and is now closely linked to computer science. An attempt to include iterative numerical methods in iteration theory would be like including the elephant in the Small Mammals House of a zoo.

Still, we mention one problem. It was noticed some time ago that by discretizing a continuous-variable problem (for instance, initial value problem) for computation purposes, chaos may appear in the solution which is not present in the rigorous solution of the original problem. Thus approximation may qualitatively falsify the solution. This phenomenon has been called "ghost dynamics" ([Ushiki 86]).

II. *Dynamics in one or two dimensions.* This important and rapidly growing field also belongs to the "hard core" of iteration theory and should be included in a survey such as this. In [Targonski 81] it was still possible to give a reasonably complete account of the field up to the time the manuscript went to the publisher. Since then, the field has grown so rapidly that I did not keep up entirely and can no longer attempt in good faith a survey.

Anyone interested in this field is well advised to look first at the book [Alsedà, Llibre, Misiurewicz 92]. In the list of references one finds at least eight more books on the subject, starting with [Collet, Eckmann 80], and a large number of papers by R. L. Adler, Ll. Alsedà, P. Blanchard, L. Block, R. Bowen, U. Burkart, A. Chenciner, P. Collet, J.-P. Eckmann, J. Franks, J. Guckenheimer, I. Gumowski, M. Hénon, M. R. Herman, P. Holmes, L. Jonker, P. E. Kloeden, A. G. Konheim, O. E. Lanford III, A. Lasota, T.-Y. Li, J. Llibre, E. N. Lorenz, A. M. McAndrew, J. Milnor, C. Mira, M. Misiurewicz, P. Mumbrú, Z. Nitecki, A. N. Sharkovsky,

C. Simó, S. Smale, J. Smítal, P. Štefan, P. D. Straffin, F. Takens, W. Thurston and many others.

All this is, of course, mainly in one dimension, but the key "key words" already appear: periodicity, chaos, topological entropy, strange attractors. For two dimensions, an introduction is provided by [Whitley 83].

For recent work applying symbolic dynamics, see [Lampreia, Sousa Ramos 91].

III. *Complex iteration.* This field would have to have a place of honour in any history of iteration theory because of the great discoveries of the XIX and early XX century. Here we hint mostly at work emerging during the past one or two decades. Complex iteration provides a particularly elegant way of dealing with certain dynamical systems in the plane. It is truly amazing what a variety of iterative behaviour can be seen even for the function $z^2 + \alpha$ as the parameter α is varied. Notions as Fatou set, Julia set, Mandelbrot set (not, of course, confined to complex iteration) arise here. For an introduction to this field see e.g. [Blanchard 84], [Douady, Hubbard 84/85]. J. Ecalle's theory of resurgent functions is an important contribution [Ecalle 81, 85].

IV. *Fractal sets.* These sets occur not only in iteration theory, but arise naturally as limit sets of splinters, boundaries of certain sets and so on. Such sets have been named "fractals" and extensively discussed by B. Mandelbrot (see [Mandelbrot 82]).

Fractals turn up also in connection with functional equations ([Dubuc 85]).

The notion of fractal seems to be related to C. Mira's concept of "frontière floue" ("vague boundary") see [Mira 79].

Also for application of fractals to computer graphics see [Barnsley 88].

For a survey of work of the Toulouse group up to 1987, see [Thibault 89].

Fractals, in particular the sets of limit points of iterative sequences in the plane can be very beautiful. This was discovered by I. Gumowski and C. Mira, who named this phenomenon "chaos esthétique" and showed an exhibition of attractive pictures in Toulouse, at the 1973 iteration conference. Others soon discovered commercial chances and now it is possible to buy posters and sets of slides showing fractal sets.

The beauty of the images can be enhanced by using different colours for the various constituents of the fractals. For a volume with such pictures see [Peitgen, Richter 86].

V. *Experimental mathematics.* This plays a part in iteration theory as in other parts of mathematics. It became feasible when high speed computing and also computer graphics became available. The "experimental results" can be used to find conjectures which then may be proved using the conventional methods of mathematics. Or, graphics may be used as "experimental proof" of statements. This is the extreme case. In general, the situation is that both "conventional" and "experimental" methods are used in a certain approach to dynamics which has an engineering flavour and which is close to applications. For an introduction see [Gumowski, Mira 80, 90₂]; see also numerous papers by C. Mira and his collaborators and students.

Concluding remark

Iteration can be considered as a field of research bordering on functional equations as well as on dynamics. This survey was prepared for publication in Aequationes Mathematicae. In order to keep the paper reasonably short, in this particular situation I leaned towards functional equations. Dynamics was emphasized less than it would have been in a more balanced treatment.

REFERENCES

[Aczél 49]   ACZÉL, J., *Einige aus Funktionalgleichungen zweier Veränderlichen ableitbare Differentialgeichungen.* Acta Sci. Math. (Szeged) *13* (1949), 179–189.

[Aczél 91]   ACZÉL, J., *Remarks on a problem of Gy. Targonski. Report of the 27th ISFE, Poland 1989.* Aequationes Math. *39* (1990), 314–315.

[Aczél, Gronau 88]   ACZÉL, J. and GRONAU, D., *Some differential equations related to iteration theory.* Canad. J. Math. *40* (1988), 695–717.

[Aczél, Gronau 88₁]   ACZÉL, J. and GRONAU, D., *Iteration, translation, commuting and differential equations.* In: Gronau, D. and L. Reich (eds), *Selected topics in functional equations.* [Grazer Math. Bericht Nr. 295], Math.-Stat. Sekt. Forschungsges. Joanneum, Graz, 1988.

[Agnes, Rasetti 88]   AGNES, C. and RASETTI, M., *Complexity, undecidability and chaos: a class of dynamical systems with fractal orbits.* In: Livi et al. (eds), *Workshop on chaos and complexity, Torino, Oct. 5–11, 1987.* World Scientific, Singapore, 1988, pp. 3–25.

[Alsedá, Llibre, Misiurewicz 92]   ALSEDÀ, LL., LLIBRE, J. and MISIUREWICZ, M., *Combinatorial dynamics and entropy in dimension one.* World Scientific, Singapore, 1992.

[Alsina et al. 89]   ALSINA, C., LLIBRE, J., MIRA, C., SIMÓ, C., TARGONSKI, GY. and THIBAULT, R., (eds), *ECIT 87, Proc. Europ. Conf. on Iteration Th., Caldes de Malavella, Sept. 20–26, 1987.* World Scientific, Singapore, 1989.

[Barnsley 88]   BARNSLEY, M., *Fractals everywhere.* Academic Press, New York, 1988.

[Bartels 91]   BARTELS, A., *Über iterative Phantomwurzeln von Abbildungen (On iterative phantom roots of mappings).* Diploma (M.Sc.) Thesis, Universität Marburg, 1991.

[Blanchard 84]   BLANCHARD, P., *Complex analytic dynamics of the Riemann sphere.* Bull. Amer. Math. Soc. *11* (1984), 95–141.

[Blažková, Chvalina 84]   BLAŽKOVÁ, R. and CHVALINA, J., *Regularity and transivity of local-automorphism semigroups of locally finite forests.* Arch. Math. (Brno) 4, Scripta Fac. Sci. Nat. USEP Brunensis *20* (1984), 183–194.

[Böttcher, Heidler 92]   BÖTTCHER, A. and HEIDLER, H., *Algebraic composition operators*. Integral Equations Operator Theory *15* (1992), 390–411.

[Bourlet 97]   BOURLET, C., *Sur certaines équations analogues aux équations différentielles (On certain equations analogous to differential equations)*. C.R. Acad. Sci. Paris *124* (1978), 1431–1433.

[Bourlet 97₁]   BOURLET, C., *Sur les transmutations (On transmutations)*. Bull. Soc. Math. France *25* (1897), 132–140.

[Burkart 82]   BURKART, U., *Zur Charakterisierung diskreter dynamischer Systeme (On characterization of discrete dynamical systems)*. Ph.D. Thesis, Universität Marburg, 1982.

[Cap 89]   CAP, C. H., *Two approaches to the iteration problem of diffeomorphisms*. In [Alsina et al. 89], pp. 139–144.

[Cap 91]   CAP, C. H., *Solving Abel, Jabotinsky and inverse ODE problems*. In [Mira et al. 91], pp. 19–28.

[Chvalina, Matoušková 84]   CHVALINA, J. and MATOUSKOVÁ, K., *Coregularity of endomorphissm monoids of unars*. Arch. Math. (Brno) 1, Scripta Fac. Sci. Nat. USEP Brunensis *20* (1984), 43–48.

[Collet, Eckmann 80]   COLLET, P. and ECKMANN, J. P., *Iterated maps on the interval as dynamical systems*. Birkhäuser, Boston, 1980.

[Deslauriers, Dubuc 91]   DESLAURIERS, G. and DUBUC, S., *Continuous iterative iteration processes*. In [Mira et al. 91], pp. 71–78.

[Douady, Hubbard 84/85]   DOUADY, A. and HUBBARD, J., *Etude dynamique de polynômes complexes (Dynamical study of complex polynomials) I, II*. [Publ. Math. Orsay 84-02 and 85-04], Univ. d'Orsay, Orsay, 1984–85.

[Dubuc 85]   DUBUC, S., *Functional equations connected with peculiar curves*. In [Liedl et al. 85], pp. 33–40.

[Ecalle 81]   ECALLE, J., *Les fonctions résurgents (On resurgent functions)*, vol. 1, 2, [Publ. Math. Orsay], Univ. d'Orsay, Orsay, 1981.

[Ecalle 85]   ECALLE, J., *Iteration and analytic classification of local diffeomorphisms of $\mathbb{C}^\nu$*. In [Liedl et al. 85], pp. 41–48.

[Farmer et al. 85]   FARMER, D. et al. (eds), *Cellular automata*. North Holland, Amsterdam, 1985.

[Ferber 85]   FERBER, R., *Zelluläre Automaten als dynamische Systeme (Cellular automata as dynamical systems)*. Diploma (M.Sc.) Thesis, Universität Marburg, 1985.

[Ferber 88]   FERBER, R., *Räumliche und zeitliche Regelmäßigkeiten zellularer Automaten (Spatial and temporal regularities of cellular automata)*. Ph.D. Thesis, Universität Marburg, 1988.

[Ferber 91]   FERBER, R., *Cellular automata are the continuous self-mappings of configuration spaces*. In [Mira et al. 91], pp. 79–85.

[Ferber et al. 91]   FERBER, R., TARGONSKI, GY. and WEITKÄMPER, J., *Fractional-time states of cellular automata*. In [Mira et al. 91], pp. 86–106.

[Förg-Rob 85]   FÖRG-ROB, W., *The Pilgerschritt transform in Lie algebras*. In [Liedl et al. 85], pp. 59–71.

[Förg-Rob 89]   FÖRG-ROB, W., *Some results on the Pilgerschritt transform*. In [Alsina et al. 89], pp. 198–204.

[Förg-Rob, Netzer 85]   FÖRG-ROB, W. and NETZER, N., *Product-integration and one-parameter subgroups of linear Lie groups*. In [Liedl et al. 85], pp. 71–82.

[Förg-Rob et al. 94]   FÖRG-ROB, W., GRONAU, D., MIRA, C., NETZER, N. and TARGONSKI, GY., (eds), *Proceedings of ECIT 92*, Batschuns (Austria), September 1992. World Scientific, Singapore, 1994.

[Gale 91] GALE, D., *Conjectures*. Math. Intelligencer *13* (1991), 53–55.

[Gardner 70] GARDNER, M., *Mathematical games*. Scientific American, Oct. 1970 and Feb. 1971.

[Graw 82] GRAW, R., *Über die Orbitstruktur stetiger Abbildungen (On the orbit structure of continuous mappings.)*. Ph.D. Thesis, Universität Marburg, 1982.

[Graw 84] GRAW, R., *Compact orbits and periodicity*. Nonlinear Anal. *8* (1984), 1473–1479.

[Gronau 91] GRONAU, D., *The Jabotinsky equations and the embedding problem*. In [Mira et al. 91], pp. 138–148.

[Gronau 91$_1$] GRONAU, D., *On the structure of the solution of the Jabotinsky equations in Banach spaces*. Zeitschr. Anal. Anw. *10* (1991), 335–343.

[Gumowski, Mira 80] GUMOWSKI, I. and MIRA, C., *Dynamique chaotique (Chaotic dynamics)*. Cepadues Editions, Toulouse, 1980.

[Gumowski, Mira 80$_1$] GUMOWSKI, I. and MIRA, C., *Recurrences and discrete dynamic systems*. [Springer Lecture Notes in Mathematics, Nr. 809], Springer, Berlin, 1980.

[Isaacs 50] ISAACS, R., *Iterates of fractional order*. Canad. J. Math. *2* (1950), 409–416.

[Jabotinsky 55] JABOTINSKY, E., *Iteration*. Ph.D. Thesis, The Hebrew University, Jerusalem, 1955.

[Jabotinsky 63] JABOTINSKY, E., *Analytic iteration*. Trans. Amer. Math. Soc. *108* (1963), 457–477.

[Krasner 38] KRASNER, M., *Une géneralisation de la notion de corps (A generalization of the notion of field)*. J. Math. Pures Appl. *17* (1938), 367–385.

[Krause 88] KRAUSE, G., Manuscript, 1988.

[Kuczma, Choczewski, Ger 90] KUCZMA, M., CHOCZEWSKI, B. and GER, R., *Iterative functional equations*. [Encyclopedia of Mathematics and its Applications, vol. 32], Cambridge University Press, Cambridge, 1990.

[Lagarias 85] LAGARIAS, J. C., *The 3x + 1 problem and its generalizations*. American Math. Monthly *92* (1985), 3–23.

[Lampreia, Sousa Ramos 91] LAMPREIA, J. P. and SOUSA RAMOS, J., *Symbolic dynamics of trimodal maps*. In [Mira et al. 91], pp. 184–193.

[Lampreia et al. 93] LAMPREIA, J. P., LLIBRE, J., MIRA, C., SOUSA RAMOS, J. and TARGONSKI, GY. (eds), *ECIT 91, Proc. Europ. Conf. on Iteration Theory, Lisbon Sep. 15–21, 1991*. World Scientific, Singapore, 1993.

[Langenberg 92] LANGENBERG, H., *Zelluläre Automaten und Iterationstheorie (Cellular automata and iteration theory)*. Diploma (M.Sc.) Thesis, Universität Marburg, 1992.

[Liedl 86] LIEDL, R., *Gruppenwertige Potenzreihen (Group valued power series)*. Anz. Österreich. Akad. Wiss. Math. Nat. Kl. 1986, No. 5, 57–58.

[Liedl, Netzer 89] LIEDL, R. and NETZER, N., *Group theoretic and differential geometric methods for solving the translation equation*. In [Alsina et al. 89], pp. 240–252.

[Liedl, Netzer 91] LIEDL, R. and NETZER, N., *Die Lösung der Translationsgleichung mittels schneller Pilgerschrittransformation (Solution of the translation equation by means of the fast Pilgerschritt transformation)*. [Grazer Ber. Nr. 314], Forschungsinst., Graz, 1991.

[Liedl, Netzer, Reitberger 81] LIEDL, R., NETZER, N. and REITBERGER, H., *Eine Methods zur Berechnung von einparametrigen Untergruppen ohne Verwendung des*

Logarithmus (A method for calculation of one-parameter subgroups without using logarithm). Österreich. Akad. Wiss., Math.-Natur. Kl. Sitzungsberg. *II* (1981), 273–284.

[Liedl, Netzer, Reitberger 82] LIEDL, R., NETZER, N. and REITBERGER, H., *Über eine Methode zur Auffindung stetiger Iterationen in Lie-Gruppen (On a method of finding continuous iteration in Lie groups)*. Aequationes Math. *24* (1982), 19–32.

[Liedl et al. 85] LIEDL, R., REICH, L. and TARGONSKI, GY. (eds), *Iteration Theory and its functional equations* (Proceedings, Schloss Hofen 1984). Springer Lecture Notes in Mathematics Nr. 1163.

[Mandelbrot 82] MANDELBROT, B. B., *The fractal geometry of nature*. W.H. Freeman, 1982.

[Miller 92] MILLER, J. B., *Square root of uppertriangular matrices*. [Analysis Paper No. 75], Dept. of Math., Monash University, Clayton, Vic., Australia, 1991.

[Mira 79] MIRA, C., *Frontière floue séparant des domaines d'attraction de deux attracteurs (Vague boundaries separating the domains of attraction of two attractors)*. C. R. Acad. Sci. Paris *299* (1979), A591–A594.

[Mira, Müllenbach 83] MIRA, C. and MÜLLENBACH, S., *Sur l'itération fractionnaire d'un endomorphisme quasratique (On fractional iteration of a quadratic endomorphism)*. C. R. Acad. Sci. Paris Sér. I. Math. *297* (1983), 369–372.

[Mira et al. 91] MIRA, C., NETZER, N., SIMÓ, C. and TARGONSKI, GY. (eds), *Proceedings of ECIT 89, European Conference on Iteration Theory, Batschuns, Austria, 10–16 Sept. 1989*. World Scientific, Singapore, 1991.

[Netzer 82] NETZER, N., *On the convergence of iterated pilgerschritt transforms*. Zeszyty Nauk. Uniw. Jagiellon. Prace Mat. *23* (1982), 91–98.

[Netzer 92] NETZER, N., *The convergence of fast Pilgerschritt transformation*. In [Förg-Rob et al. 92].

[Netzer, Liedl 91] NETZER, N. and LIEDL, R., *Fast Pilgerschritt transformation*. In [Mira et al. 91], pp. 279–293.

[Netzer, Reitberger 82] NETZER, N. and REITBERGER, H., *On the convergence of Pilgerschritt transformations in nilpotent Lie groups*. Publ. Math. Debrecen *29* (1982), 309–314.

[Neuman 82] NEUMAN, F., *Simultaneous solutions of a system of Abel equations and differential equations with several derivations*. Czechoslovak Math. J. *32* (1982) 488–494.

[Neuman 89] NEUMAN, F., *On iteration groups of certain functions*. Arch. Math. (Brno) *25* (1989), 185–194.

[Von Neumann 66] VON NEUMANN, J., *Theory of self-reproducing automata*. University of Illinois, Urbana-London, 1966.

[Peitgen, Richter 86] PEITGEN, H. O. and RICHTER, P. H., *The beauty of fractals*. Springer, Berlin, 1986.

[Peschl, Reich 71] PESCHL, E. and REICH, L., *Eine Linearisierung kontrahierender biholomorpher Abbildungen und damit zusammenhängender analytischer Differentialgleichungssysteme (A linearization method for contractive biholomorphic maps and for related systems of analytic differential equations)*. Monatsh. Math. *75* (1971), 153–162.

[Reich 71] REICH, L., *Über analytische Iteration linearer und kontrahierender biholomorpher Abbildungen (On analytic iteration of linear and contractive biholomorphic mappings)*. [Bericht Nr. 42] Ges. Math. Datenverarb., Bonn, 1971.

[Reich 79]   REICH, L. (with the participation of J. Schwaiger), *Analytische und fraktionelle Iteration formal-biholomorpher Abbildungen (Analytic and fractional iteration of formal biholomorphic maps)*. In *Jahrbuch Überblicke Mathematik 1979*, Bibliographisches Institut, Mannheim, 1979, pp. 123–144.

[Reich 85]   REICH, L., *On a differential equation arising in iteration theory in rings of formal power series in one variable*. In [Liedl et al. 85], pp. 135–148.

[Reich 88]   REICH, L., *On families of commuting formal power series*. In *Selected topics in functional equations* [Grazer Math. Bericht Nr. 294] Math-Stat. Sekkt. Forsch Ges. Joanneum, Graz, 1988, pp. 1–18.

[Reich 89]   REICH, L., *Die Differentialgleichungen von Aczél–Jabotinsky, von Briot–Bouquet und maximale Familien vertauschbarer Potenzreihen (The differential equations of Aczél–Jabotinsky, of Briot–Bouquet and maximal families of commuting power series)*. In *Complex methods on partial differential equations*. [Math. Res. Vol. 53]. Akademie Verlag, Berlin, 1989, pp. 137–150.

[Reich 91]   REICH, L., *On the embedding problem for formal power series with respect to the Aczél–Jabotinsky equations*. In [Mira et al. 91], pp. 294–304.

[Reich 92]   REICH, L., *On the local distribution of iterable power series transformation in one indeterminate*. In: *Functional analysis III, Proc. Postgrad School and Conf., Dubrovnik, Oct. 29–Nov. 2* (D Brutkovic et al. eds). [Aarhus Univ. Various Publications Series Nr. 40], Univ., Aarhus, 1992.

[Reich 93]   REICH, L., *On power series transformations in one indeterminate having iterative roots of a given order and with given multiplier*. In [Lampreia et al. 93], pp. 210–216.

[Reich, Schwaiger 80]   REICH, L. and SCHWAIGER J., *Linearisierung formal-biholomorpher Abbildungen und Iterationsprobleme*. Aequationes Math. 20 (1980), 224–243.

[Riggert 75]   RIGGERT, G., *n-te iterative Wurzeln von beliebigen Abbildungen (n-th iterative roots of arbitrary sets)*. In *Report of the 1975 International Symposium on Functional Equations*. Aequationes Math. 15 (1977), 288.

[Robert 86]   ROBERT, F., *Discrete iterations*. Springer, Berlin, 1986.

[Schleiermacher 93] SCHLEIERMACHER, A., *On a theorem of Marc Krasner about invariant relations*. In [Lampreia et al. 93], pp. 230–240.

[Schwaiger 89]   SCHWAIGER, J., *Phantom roots and phantom iterates of formal power series in one variable*. In [Alsina et al. 89], pp. 313–323.

[Schwaiger 91]   SCHWAIGER, J., *On polynomials having different types of roots*. In [Mira et al. 91], pp. 315–319.

[Schweizer, Sklar 88] SCHWEIZER, B. and SKLAR, A., *Invariants and equivalence classes of polynomials under linear conjugacy*. In *Contributions to general algebra, No. 6*. Hölder–Pinchler–Tempsky, Vienna and Teubner, Stuttgart, 1988.

[Schweizer, Sklar 90] SCHWEIZER, B. and SKLAR, A., *The baker's transformation is not embeddable*. Found. of Phys. 20 (1990), 873–897.

[Simon 93]   SIMON, K., *Hausdorff dimensions for certain near-hyperbolic maps*. In [Lampreia et al. 93], pp. 253–261.

[Sklar 69]   SKLAR, A., *Canonical decompositions, stable functions, and fractional iterates*. Aequationes Math. 3 (1969), 118–129.

[Sklar 87]   SKLAR, A., *The structure of one-dimensional flows with continuous trajectories*. Rad. Mat. 3 (1987), 111–142.

[Skornjakov 77]   SKORNJAKOV, L. A., *Unars*. In *Universal algebra*. [Coll. Mat. Soc. János Bolyai 29], J. Bolyai Math. Soc. Budapest, 1977, pp. 735–743.

[Smajdor 85]   SMAJDOR, A., *Iteration of multi-valued functions*. [Prace Nauk. Uniw. Sląsk. Katowic. No. 759], Silesian Univ. Katowice, 1985.

[Smajdor 89]   SMAJDOR, A., *One-parameter families of set-valued contractions*. In [Alsina et al. 89].

[Smajdor 93]   SMAJDOR, A., *Almost-everywhere set-valued semigroups*. In [Lampreia et al. 93], pp. 262–272.

[Smítal 88]   SMÍTAL, J., *On functions and functional equations*. Adam Hilger, Bristol–Philadelphia, 1988.

[Snowden, Howie 82] SNOWDEN, M. and HOWIE, J. M., *Square roots in finite full transformation semigroups*. Glasgow Math. J. *23* (1982), 137–149.

[Targonski 67]   TARGONSKI, GY., *Seminar on functional operators and equations*. [Springer Lecture Notes in Mathematics, No. 33], Springer, Berlin, 1967.

[Targonski 81]   TARGONSKI, GY., *Topics in iteration theory*. Vandenhoeck & Ruprecht, Göttingen–Zürich, 1981.

[Targonski 84]   TARGONSKI, GY., *New directions and open problems in iteration theory*. [Grazer Math. Bericht No. 229], Math. Stat. Sekt. Forschungszentrum, Graz, 1984.

[Targonski 84₁] TARGONSKI, GY., *Phantom iterates of continuous functions*. In [Liedl et al. 85], pp. 196–202.

[Targonski 89]   TARGONSKI, GY., *Iteration theory and functional analysis*. In [Alsina et al. 89], pp. 74–93.

[Targonski 90]   TARGONSKI, GY., *On composition operators*. Zeszyty Nauk. Politechn. Śląsk. Ser. Mat.-fiz. *64* (1990).

[Targonski 91]   TARGONSKI, GY., *Problem 25*, In *Report of the 27th ISFE, Poland 1989*. Aequationes Math. *39* (1990), 313–314.

[Targonski 93]   TARGONSKI, GY., *On a class of phantom fractional iterates*. In [Lampreia et al. 93], pp. 295–301.

[Targonski 94]   TARGONSKI, GY., *Phantom iterates and Liedl's Pilgerschritt transformation*. In [Förg-Rob et al. 94].

[Targonski, Zdun 85] TARGONSKI, GY. and ZDUN, M. C., *Generators and co-generators of substitution semigroups*. Ann. Math. Sil. *1 (13)* (1985), 169–174.

[Targonski, Zdun 87] TARGONSKI, GY. and ZDUN, M. C., *Substitution operators on Lᵖ-spaces and their semigroups*. [Grazer Math. Bericht No. 283], Math.-Stat. Sekt. Forsch. Ges. Joanneum, Graz, 1987.

[Thibault 89]   THIBAULT, R., *Some results obtained in Toulouse on dynamical systems*. In [Alsina et al. 89], pp. 94–112.

[Ushiki 86]   USHIKI, SH., *Chaotic Phenomena and Fractal Objects in Numerical Analysis*. In: Nishida, T. et al. (eds), *Patterns and waves — qualitative analysis of nonlinear differential equations*. [Stud. Math. Appl.], North Holland, Amsterdam, 1986, pp. 221–258.

[Wagon 85]   WAGON, S., *The Collatz problem*. Math. Intelligencer *7* (1985), 72–76.

[Weitkämper 85] WEITKÄMPER, J., *Embeddings in iteration groups and semigroups with nontrivial units*. Stochastica *7* (1983), 175–195.

[Whitley 83]   WHITLEY, D., *Discrete dynamical systems in dimensions one and two*. Bull. London Math. Soc. *15* (1983), 177–217.

[Zdun 79]   ZDUN, M. C., *Continuous and differentiable iteration semigroups*. [Prace Nauk. Uniw. Śląsk. Katowic. No. 308], Silesian Univ., Katowice, 1979.

[Zdun 85]   ZDUN, M. C., *On embedding of the circle in a continuous flow* in [Liedl et al. 85], pp. 218–231.

[Zdun 85₁]   ZDUN, M. C., *Regular fractional iteration*. Aequationes Math. *28* (1985), 73–79.

[Zdun 88]   ZDUN, M. C., *Note on commutable functions*. Aequationes Math. *36* (1988), 153–164.

[Zdun 89]   ZDUN, M. C., *On continuity of iteration semigroups on metric spaces*.
                     Annal. Soc. Math. Polon. *29* (1989), 113–116.
[Zdun 89₁]   ZDUN, M. C., *On Cʳ iteration groups*. In [Alsina et al. 89], pp. 373–381.
[Zdun 89₂]   ZDUN, M. C., *On simultaneous Abel equations*. Aequationes Math. *38*
                     (1989), 163–177.
[Zdun 90]   ZDUN, M. C., *On quasi-continuous iteration semigroups and groups of real
                     functions*. Colloq. Math. *58* (1990), 281–289.
[Zdun 91]   ZDUN, M. C., *The structure of iteration groups of continuous functions*.
                     Aequationes Math. *46* (1993), 19–37.
[Zdun 91₁]   ZDUN, M. C., *On continuous iteration groups of fixed point free mappings
                     in $\mathbb{R}^n$ space*. In [Mira et al. 91], pp. 362–368.
[Zdun 93]   ZDUN, M. C., *Some remarks on the iterates of commuting functions*. In
                     [Lampreia et al. 93], pp. 336–342.

Fachbereich Mathematik,
Universität Marburg,
D-35032 Marburg,
Germany.

Aequationes Mathematicae **50** (1995) 73–94
University of Waterloo

0001–9054/95/020073–22 $1.50 + 0.20/0
© 1995 Birkhäuser Verlag, Basel

On the representation of integers as sums of triangular numbers

Ken Ono, Sinai Robins, and Patrick T. Wahl

Summary. In this survey article we discuss the problem of determining the number of representations of an integer as sums of triangular numbers. This study yields several interesting results. If $n \geq 0$ is a non-negative integer, then the nth triangular number is $T_n = n(n+1)/2$. Let k be a positive integer. We denote by $\delta_k(n)$ the number of representations of n as a sum of k triangular numbers. Here we use the theory of modular forms to calculate $\delta_k(n)$. The case where $k = 24$ is particularly interesting. It turns out that, if $n \geq 3$ is odd, then the number of points on the 24 dimensional Leech lattice of norm $2n$ is $2^{12}(2^{12} - 1)\delta_{24}(n-3)$. Furthermore the formula for $\delta_{24}(n)$ involves the Ramanujan $\tau(n)$-function. As a consequence, we get elementary congruences for $\tau(n)$. In a similar vein, when p is a prime, we demonstrate $\delta_{24}(p^k - 3)$ as a Dirichlet convolution of $\sigma_{11}(n)$ and $\tau(n)$. It is also of interest to know that this study produces formulas for the number of lattice points inside k-dimensional spheres.

1. Introduction

Representations of non-negative integers by quadratic forms as sums of squares have a long history [10]. For example, if $k \geq 1$ is a positive integer, then the number of representations of n as a sum of k squares, denoted by $r_k(n)$, has received considerable attention from Rankin [20]. To study $r_k(n)$, he used the classical theta function defined by

$$\Theta(q) = \sum_{n=-\infty}^{\infty} q^{n^2} = 1 + 2q + 2q^4 + 2q^9 + 2q^{16} + \cdots.$$

Consequently, the values of $r_k(n)$ are the formula coefficients of the power series

$$\Theta^k(q) = \sum_{n \geq 0} r_k(n)q^n.$$

AMS (1991) subject classification: Primary 11F11, 11F12, 11F37, Secondary 11P81.

Manuscript received August 26, 1993 and, in final form, December 6, 1993.

Fortunately, it is well known that $\Theta(q)$ is a modular form of weight $\frac{1}{2}$ on $\Gamma_0(4)$. It now follows that the modular form theory of $\Theta^k(q)$ defines $r_k(n)$. When k is odd, these calculations can be troublesome since $\Theta^k(q)$ is a modular form of half integral weight. Here we apply these classical methods to the representations of integers as sums of triangular numbers. Much of what follows is a special case of the problem of representations of numbers as sums of figurate numbers. We will show that the general case follows from the study of generalized Dedekind η-functions with little complication [22].

First we define the triangular numbers.

DEFINITION. If n is a non-negative integer, then the triangular number T_n is defined by

$$T_n = \frac{n(n+1)}{2}.$$

Note that geometrically T_n is equal to the number of nodes that complete an equilateral triangle with sidelength n. Here are the first few triangular numbers:

$$T_0 = 0 \qquad T_1 = 1 \qquad T_2 = 3 \qquad T_3 = 6 \qquad T_4 = 10.$$

If $k \geq 1$ is a positive integer, then let $\delta_k(n)$ denote the number of representations of n as a sum of k triangular numbers. We calculate $\delta_k(n)$ for several values of k using modular form theory.

Incidentally, Gauss's famous *Eureka* theorem states that every non-negative integer is represented as a sum of three triangular numbers. In our notation this says that, if $n \geq 0$, then $\delta_3(n) > 0$. The reader may consult [1] for a discussion of this theorem.

Now we give some preliminaries from the theory of modular forms. Let $N \geq 1$ be a rational integer. Then we define the following congruence subgroups of $SL_2(\mathbb{Z})$. Let A denote the following matrix with integer entries in $SL_2(\mathbb{Z})$:

$$A = \begin{pmatrix} a & b \\ c & d \end{pmatrix}.$$

DEFINITION. The most common *congruence subgroups of level* N are defined by

(i) $\Gamma_0(N) = \{A \in SL_2(\mathbb{Z}) \,|\, c \equiv 0 \bmod N\}$
(ii) $\Gamma_1(N) = \{A \in SL_2(\mathbb{Z}) \,|\, a \equiv d \equiv 1 \bmod N \text{ and } c \equiv 0 \bmod N\}$
(iii) $\Gamma(N) = \{A \in SL_2(\mathbb{Z}) \,|\, a \equiv d \equiv 1 \bmod N \text{ and } b \equiv c \equiv 0 \bmod N\}$.

Let χ be a Dirichlet character mod N and $k \in \mathbb{Z}^+$ satisfying $\chi(-1) = (-1)^k$. Let $A \in SL_2(\mathbb{Z})$ act on H, the complex upper half plane, by the linear fractional transformation

$$A\tau = \frac{a\tau + b}{c\tau + d}.$$

Let $f(\tau)$ be a holomorphic function on H such that

$$f(A\tau) = \chi(d)(cr + d)^k f(\tau)$$

for all $A \in \Gamma_0(N)$ and all $\tau \in H$. We call such $f(\tau)$ *a modular form of weight k* and character χ on $\Gamma_0(N)$. If $f(\tau)$ is holomorphic (resp. vanishes) at the *cusps* of $\Gamma_0(N)$ then $f(\tau)$ is a holomorphic modular form (resp. cusp form). The holomorphic modular forms and cusp forms of weight k and character χ form finite dimensional $\mathbb{C}$-vector spaces. These spaces are denoted by $M_k(\Gamma_0(N), \chi)$ and $S_k(\Gamma_0(N), \chi)$, respectively. It is well known that $M_k(\Gamma_0(N), \chi)$ is the direct sum of $S_k(\Gamma_0(N), \chi)$ and forms known as Eisenstein series. When $k = \lambda + \frac{1}{2}$ with $\lambda \in \mathbb{N}$, there is a similar theory of modular forms with half-integral weight [12].

Since the transformation $\tau \to \tau + 1$ is in $\Gamma_0(N)$, a holomorphic modular form $f(\tau)$ admits a Fourier expansion at the point at infinity in the uniformizing variable $q = e^{2\pi i \tau}$

$$f(\tau) = \sum_{n=0}^{\infty} a(n)q^n.$$

Understanding the arithmetic nature of these coefficients $a(n)$ has been a major topic in number theory; their behavior is related to quadratic forms, elliptic curves, integral lattices, the splitting of prime ideals in number fields etc. . . It is of interest to know that the Fourier coefficients of Eisenstein series are determined by generalized divisor functions.

There are natural linear transformations, the Hecke operators, which act on Fourier expansions of modular forms preserving $M_k(\Gamma_0(N), \chi)$ and $S_k(\Gamma_0(N), \chi)$. If p is prime, then the Hecke operator T_p is defined by

$$f|T_p = \sum_{n=0}^{\infty} a(pn)q^m + \chi(p)p^{k-1} \sum_{n=0}^{\infty} a(n)q^{pn}.$$

Note that if $p|N$, then $\chi(p) = 0$, so T_p reduces to the dissection operator U_p defined by

$$f|U_p = \sum_{n=0}^{\infty} a(pn)q^n.$$

For a thorough treatment of the theory of modular forms the reader should consult [11], [12], [15], [23] or [28].

2. The generating function

To carry out our study, we will use the generating function Ψ, defined by

$$\Psi(q) = \sum_{n=0}^{\infty} q^{T_n} = 1 + q + q^3 + q^6 + q^{10} + \cdots .$$

It is easy to see that if $k \geq 1$ then

$$\Psi^k(q) = \sum_{n=0}^{\infty} \delta_k(n)q^n.$$

We will see that $\Psi(q)$ is essentially a quotient of Dedekind η-functions. The Dedekind η-function is a modular form of weight $\frac{1}{2}$ that is defined by the infinite product

$$\eta(r) = q^{1/24} \prod_{n=1}^{\infty} (1 - q^n) \qquad \text{where } q = e^{2\pi i \tau}.$$

Products and quotients of this function have been studied extensively because of their applications to combinatorics and representations of the symmetric group [1], [2], [7], [8], [18], [21].

We are interested in $q\Psi(q^8)$; it is the Fourier expansion for the weight $\frac{1}{2}$ modular form

$$\frac{\eta^2(16\tau)}{\eta(8\tau)} = q + q^9 + q^{25} + q^{49} + \cdots .$$

Using the Serre-Stark Basis theorem [27], it turns out that $\eta^2(16\tau)/\eta(8\tau)$ is the theta series

$$\frac{\eta^2(16\tau)}{\eta(8\tau)} = \sum_{\substack{n \geq 1 \\ n \text{ odd}}} q^{n^2}.$$

Here we use this result to derive an infinite product representation for $\Psi(q)$:

$$q^{-1/8} \frac{\eta^2(2\tau)}{\eta(\tau)} = \sum_{\substack{n \geq 0 \\ n \text{ odd}}} q^{(n^2 - 1)/8} = \sum_{n=0}^{\infty} q^{T_n} = \Psi(q).$$

This proves the following proposition.

PROPOSITION 1. *If T_n is the nth triangular number, then*

$$\Psi(q) = \prod_{n=1}^{\infty} \frac{(1-q^{2n})^2}{(1-q^n)} = \sum_{n=0}^{\infty} q^{T_n}.$$

Furthermore, we establish a simple relationship between square and triangular representations:

PROPOSITION 2. *If $\delta_k(n)$ is the number of representations of n as a sum of k triangular numbers and $q_k(n)$ is the number of representations of n as a sum of k odd squares then*

$$\delta_k(n) = q_k(8n + k).$$

Proof. This follows by rewriting a representation of n as a sum of k triangular numbers in the following way:

$$n = \sum_{i=1}^{k} \frac{x_i(x_i + 1)}{2} \Leftrightarrow 8n = \sum_{i=1}^{k} (4x_i^2 + 4x_i)$$

$$\Leftrightarrow 8n + k = \sum_{i=1}^{k} (2x_i + 1)^2. \qquad \square$$

3. Formulae for some $\delta_k(n)$

In this section we compute formulae for $\delta_k(n)$ for various values of k. When $k = 2$ or 3, this reduces to calculating $r_2(8n + 2)$ and $r_3(8n + 3)$. For other values of k we apply classical modular form theory.

The case $k = 2$:

By Proposition 2 we know that $\delta_2(n) = q_n(8n + 1)$. It is easy to see that, if $\alpha, \beta \in \mathbb{Z}$ are solutions to

$$\alpha^2 + \beta^2 \equiv 2 \bmod 8,$$

then α and β are necessarily odd. Consequently, we obtain

$$\delta_2(n) = q_2(8n + 2) = \frac{1}{4} r_2(8n + 2).$$

The scalar $\frac{1}{4}$ compensates for the 4 possible choices of sign that are counted with multiplicity in $r_2(n)$. Here we state the well-known formula for $r_2(n)$ [10, p.15].

THEOREM 1. *Denote the number of divisors of n by d(n), and write $d_a(n)$ for the number of divisors d of n with $d \equiv a$ mod 4. Let $n = 2^l n_1 n_2$, where the prime factorizations of n_1 and n_2 are*

$$n_1 = \prod_{p \equiv 1 \bmod 4} p^r \quad and \quad n_2 = \prod_{q \equiv 3 \bmod 4} q^s.$$

If any of the exponents s are odd, then $r_2(n) = 0$. If all s are even, then

$$r_2(n) = 4d(n_1) = 4(d_1(n) - d_3(n)).$$

As a simple corollary we get

COROLLARY 1. *Let $8n + 2 = 2^l n_1 n_2$, where*

$$n_1 = \prod_{p \equiv 1 \bmod 4} p^r \quad and \quad n_2 = \prod_{q \equiv 3 \bmod 4} q^s.$$

If any of the exponents s are odd, then $\delta_2(n) = 0$. If all of the exponents s are even, then

$$\delta_2(n) = d(n_1) = d_1(8n + 2) - d_3(8n + 2).$$

The case $k = 3$:

Here we make use of the identity

$$\delta_3(n) = q_3(8n + 3).$$

It is easy to verify that, if $\alpha, \beta, \gamma \in \mathbb{Z}$ are solutions of

$$\alpha^2 + \beta^2 + \gamma^2 \equiv 3 \bmod 8,$$

then α, β and γ are all odd. Consequently, we obtain the identity

$$\delta_3(n) = q_3(8n + 3) = \frac{1}{8} r_3(8n + 3).$$

Here the scalar $\frac{1}{8}$ compensates for the 8 choices of sign that are counted with multiplicity in $r_3(n)$. Again we state a classical result [10, p. 53] that gives a formula for $r_3(n)$.

THEOREM 2. *Let $[x]$ be the greatest integer function and let (r/n) be the usual Jacobi symbol.*
If $n \equiv 1 \bmod 4$ then define $R_3(n)$ by

$$R_3(n) = 24 \sum_{r=1}^{[n/4]} \left(\frac{r}{n}\right).$$

If $n \equiv 3 \bmod 4$ then define $R_3(n)$ by

$$R_3(n) = 8 \sum_{r=1}^{[n/2]} \left(\frac{r}{n}\right).$$

With these definitions, we have

$$r_3(n) = \sum_{d^2 \mid n} R_3\left(\frac{n}{d^2}\right).$$

As a corollary, we obtain

COROLLARY 2. *Given the above notation above we have*

$$\delta_3(n) = \frac{1}{2} \sum_{d^2 \mid (8n+3)} R_3\left(\frac{8n+3}{d^2}\right).$$

The case $k = 4$:

It is easy to see that $q\Psi^4(q^2)$ is the weight 2 modular form on $\Gamma_0(4)$ defined by

$$q\Psi^4(q^2) = \frac{\eta^8(4\tau)}{\eta^4(2\tau)} = q\left[\prod_{n=1}^{\infty} \frac{(1-q^{4n})^2}{(1-q^{2n})}\right]^4.$$

Its Fourier expansion is

$$q\Psi^4(q^2) = \sum_{n=0}^{\infty} \delta_4(n)q^{2n+1} = q + 4q^3 + 6q^5 + 8q^7 + 13q^9 + \cdots.$$

Incidentally note that, if $k \geq 1$ is a positive integer, then the kth power of $q\Psi^4(q^2)$ defines $\delta_{4k}(n)$ by

$$q^4\Psi^{4k}(q^2) = \frac{\eta^{8k}(4\tau)}{\eta^{4k}(2\tau)} = \sum_{n=0}^{\infty} \delta_{4k}(n)q^{2n+k}.$$

Since spaces of modular forms are finite dimensional, one easily determines that two modular forms with the same level and weight are equal if their Fourier expansions agree for sufficiently many terms. We only need to check the first $k[SL_2(\mathbb{Z}):\Gamma']/(12)$ terms where k is the weight and Γ' is the relevant congruence subgroup [12] [28].

It turns out that $q\Psi^4(q^2)$ is the Eisenstein series on $\Gamma_0(4)$, defined by

$$q\Psi^4(q^2) = \sum_{n=0}^{\infty} \sigma_1(2n+1)q^{2n+1},$$

where $\sigma_k(m) = \sum_{d|m} d^k$ is the standard divisor function. We obtain

THEOREM 3. *If $\delta_4(n)$ is the number of representations of n as a sum of 4 triangular numbers, then*

$$\delta_4(n) = \sigma_1(2n+1).$$

This result was known to Legendre [6], [14].

As a consequence of the multiplicativity of $\sigma_1(n)$, we establish the following interesting multiplicity property for $\delta_4(n)$:

$$\delta_4(m)\delta_4(n) = \delta_4(2mm + m + n) \qquad \text{when } (2m+1, 2n+1) = 1.$$

Incidentally, $q\Psi^4(q^2)$ is a basis element for the spaces of modular forms of level 4. Here we state a well known fact [12, p. 184] that determines a polynomial basis for modular forms on $\Gamma_0(4)$ of either integral or half integral weight.

PROPOSITION 3. *Define $\Theta(\tau)$ and $q\Psi^4(q^2)$ as above. $\Theta(\tau)$ has weight $\frac{1}{2}$ and $q\Psi^4(q^2)$ has weight 2. If $k \in \mathbb{Z}$, then $M_{k/2}(\Gamma_0(4))$ is the space of all isobaric polynomials in $\mathbb{C}[\Theta(\tau), q\Psi^4(q^2)]$ with weight $k/2$.*

This proposition says that all modular forms on $\Gamma_0(4)$ have Fourier expansions at $i\infty$ coming from triangular numbers and squares. Later in this paper we define generating functions for the generic figurate numbers as weight $k = \frac{1}{2}$ modular forms. Is there some general result which allows us to determine when the generating functions of a set of figurate numbers defines a polynomial basis for all modular forms (i.e. integral and half-integral weight) of given level and character?

The case $k = 6$:

We consider the modular form $q^3 \Psi^6(q^4) \in M_3(\Gamma_0(8))$.

$$q^3 \Psi^6(q^4) = \frac{\eta^{12}(8\tau)}{\eta^6(4\tau)} = q^3 \left[\prod_{n=1}^{\infty} \frac{(1 - q^{8n})^2}{(1 - q^{4n})} \right]^2 = q^3 \sum_{n=0}^{\infty} \delta_6(n)q^{4n} = \sum_{n=0}^{\infty} \delta_6(n)q^{4n+3}.$$

The first few terms of the Fourier expansion of $q^3 \Psi^6(q^4)$ are

$$q^3 \Psi^6(q^4) = q^3 + 6q^7 + 15q^{11} + 26q^{15} + 45q^{19} + 66q^{23} + 82q^{27} + \cdots.$$

Define χ, the Dirichlet character mod 4, by

$$\chi(1) = 1, \qquad \chi(3) = -1,$$

define the generalized divisor function $\sigma_{2,\chi}(n)$ by

$$\sigma_{2,\chi}(n) = \sum_{d|n} \chi(d)d^2,$$

and define $G_{2,\chi}(\tau)$, a weight 3 Eisenstein series on $\Gamma_0(8)$, by

$$G_{2,\chi}(\tau) = \sum_{n=0}^{\infty} \sigma_{2,\chi}(4n + 3)q^{4n+3}.$$

It turns out that our generating function $q^3 \Psi^6(q^4)$ satisfies

$$q^3 \Psi^6(q^4) = -\frac{1}{8} G_{2,\chi}(\tau).$$

We obtain

THEOREM 4: *If $\delta_6(n)$ is the number of representations of n as sum of 6 triangular numbers, then*

$$\delta_6(n) = -\frac{1}{8} \sigma_{2,\chi}(4n + 3).$$

Note that this theorem naturally implies that $\delta_6(n)$ satisfies some interesting multiplicative properties.

The case $k = 8$:

Here we consider the weight 4 modular form $q^2\Psi^8(q^2)$ on $\Gamma_0(4)$ defined by

$$q^2\Psi^8(q^2) = \frac{\eta^{16}(4)}{\eta^8(2\tau)} = \sum_{n=0}^{\infty} \delta_8(n)q^{2n+2}.$$

Here are the first few terms of the Fourier expansion of $q^2\Psi^8(q^2)$

$$q^2\Psi^8(q^2) = q^2 + 8q^4 + 28q^6 + 64q^8 + 126q^{10} + 224q^{12} + \cdots.$$

Here we define $E(\tau)$, a well known weight 4 Eisenstein series on $\Gamma_0(2)$,

$$E(\tau) = \sum_{n=1}^{\infty} \sigma_3^{\#}(n)q^n,$$

where

$$\sigma_3^{\#}(n) = \sum_{\substack{d|n \\ n/d \text{ odd}}} d^3.$$

Replacing q by q^2 gives us an Eisenstein series $E(2\tau)$ on $\Gamma_0(4)$ with Fourier expansion

$$E(2\tau) = \sum_{n=1}^{\infty} \sigma_3^{\#}(n)q^{2n}.$$

By equating Fourier coefficients we find that

$$q^2\Psi^8(q^2) = E(2\tau).$$

This proves

THEOREM 5. *If $\delta_8(n)$ is the number of representations of n as a sum of 8 triangular numbers, then*

$$\delta_8(n) = \sigma_3^{\#}(n+1).$$

As a consequence of the multiplicativity of $\sigma_3^{\#}(n)$, we get following multiplicative property for $\delta_8(n)$.

COROLLARY 3. *If $\delta_8(n)$ is the number of representations of n as a sum of 8 triangular numbers and if $(a + 1, b + 1) = 1$, then*

$$\delta_8(a)\delta_8(b) = \delta_8((a + 1)(b + 1) - 1).$$

The case $k = 10$:

In this case we are interested in the weight 5 modular form on $\Gamma_0(8)$ given by $q^5\Psi^{10}(q^4)$

$$q^5\Psi^{10}(q^4) = \frac{\eta^{20}(8\tau)}{\eta^{10}(4\tau)} = \sum_{n=0}^{\infty} \delta_{10}(n)q^{4n+5}.$$

Unfortunately, $q^5\Psi^{10}(q^4)$ is not an Eisenstein series; it is a linear combination of an Eisenstein series and a cusp with complex multiplication [24].

Let $F(\tau)$ be the modular form with complex multiplication by $Q(i)$ defined as

$$F(\tau) = \eta^4(\tau)\eta^2(2\tau)\eta^4(4\tau) = \sum_{n=1}^{\infty} a(n)q^n = q - 4q^2 + 16q^4 - 14q^5 - 64q^8 + \cdots.$$

It turns out that

$$q^5\Psi^{10}(q^4) = \frac{1}{640}\left[G_{4,\chi}(\tau) - 2F(\tau) - \frac{1}{4}(F(\tau)|T_2) \right]. \tag{1}$$

Here $G_{4,\chi}(\tau)$ is the Eisenstein series defined by

$$G_{4,\chi}(\tau) = \sum_{n \equiv 1 \bmod 4} \sigma_{4,\chi}(n)q^n$$

and

$$\sigma_{4,\chi}(n) = \sum_{d|n} \chi(d)\, d^4.$$

The character χ is the same Dirichlet character mod 4 which occurred when $k = 6$. Now we can state the explicit formula for $\delta_{10}(n)$.

THEOREM 6. *If $\delta_{10}(n)$ is the number of representations of n as a sum of 10 triangular numbers then*

$$\delta_{10}(n) = \frac{1}{640}[\sigma_{4,\chi}(4n+5) - a(4n+5)].$$

Proof. By equation (1), we obtain

$$\delta_{10}(n) = \frac{1}{640}\left[\sigma_{4,\chi}(4n+5) - 2a(4n+5) - \frac{1}{4}a(8n+10)\right].$$

Since 2 divides the level, $T_2 = U_2$ and one easily verifies that

$$F(\tau)|U_2 = -4F(\tau).$$

This imples that $-4a(n) = a(2n)$. Consequently, we obtain

$$\delta_{10}(n) = \frac{1}{640}[\sigma_{4,\chi}(4n+5) - a(4n+5)]. \qquad \square$$

Incidentally, since all forms with compelx multiplication are lacunary, (i.e. the arithmetic density of their non-zero Fourier coefficients is 0) this theorem implies that $\delta_{10}(n) = \frac{1}{640}\sigma_{4,\chi}(4n+5)$ almost always. For lacunarity, the reader should consult [9], [24], [25]

The case $k = 12$:

Here we consider the weight 6 modular form on $\Gamma_0(4)$ with Fourier expansion $q^3\Psi^{12}(q^2)$,

$$q^3\Psi^{12}(q^2) = \frac{\eta^{24}(4\tau)}{\eta^{12}(2\tau)} = \sum_{n=0}^{\omega} \delta_{12}(n)q^{2n+3}.$$

Here are the first few terms of the Fourier expansion of $q^3\Psi^{12}(q^2)$:

$$q^3\Psi^{12}(q^2) = q^3 + 12q^5 + 66q^7 + 232q^9 + 627q^{11} + 1452q^{13} + \cdots.$$

The space of cusp forms $S_6(\Gamma_0(4))$ is 1 dimensional and is spanned by the η-product

$$\eta^{12}(2\tau) = q - 12q^3 + 54q^5 - 88q^7 - 99q^9 + \cdots.$$

Our form $q^3\Psi^{12}(q^2)$ satisfies the following identity:

$$q^3\Psi^{12}(q^2) = \frac{1}{256}[E(\tau) - \eta^{12}(2\tau)]$$

where $E(\tau) = \sum_{n=0}^{\infty} \sigma_5(2n+1)q^{2n+1}$. Consequently, we have proved the following theorem.

THEOREM 7. *If* $\eta^{12}(2\tau) = \sum_{n=1}^{\infty} a(n)q^n$ *and* $\delta_{12}(n)$ *is the number of representations of* n *as a sum of* 12 *triangular numbers then*

$$\delta_{12}(n) = \frac{\sigma_5(2n+3) - a(2n+3)}{256}.$$

As a simple consequence of this formula, we obtain the following mod 256 congruence for $a(n)$, the Fourier coefficients of $\eta^{12}(2\tau)$:

$$a(2n+1) \equiv \sigma_5(2n+1) \bmod 256.$$

4. The Ramanujan function $\tau(n)$, the Leech lattice, and $\delta_{24}(n)$

In this section we derive a formula for $\delta_{24}(n)$, the number of representations of n as a sum of 24 triangular numbers. It turns out that for odd n we obtain an interesting relation between $\delta_{24}(n-3)$ and N_{2n}, the number of lattice points on the Leech lattice with norm $2n$ [3, p. 131]. As a corollary to the formula for $\delta_{24}(n)$, we get interesting congruences for the Ramanujan $\tau(n)$ function. Recall that $\tau(n)$ is defined to be the nth Fourier coefficient of the unique normalized weight 12 cusp form on $SL_2(\mathbb{Z})$, $\Delta(\tau) = \eta^{24}(\tau)$.

The congruence behavior of $\tau(n)$ has been studied extensively by Rankin, Serre and Swinnerton-Dyer [16], [17], [19], [26], [29], [30]. The theory of l-adic Galois representations due to Deligne and Serre [4], [5] provide a theoretical interpretation of these congruences. Here we recall the congruences for $\tau(n) \bmod 256$ and mod 691:

$$\tau(2n+1) \equiv \sigma_{11}(2n+1) \bmod 256$$

and

$$\tau(n) \equiv \sigma_{11}(n) \bmod 691.$$

Kolberg [13] extended the mod 256 congruence by proving

$$\tau(8n + 1) \equiv \sigma_{11}(8n + 1) \bmod 2^{11},$$

$$\tau(8n + 3) \equiv 1217\sigma_{11}(8n + 3) \bmod 2^{13},$$

$$\tau(8n + 5) \equiv 1537\sigma_{11}(8n + 7) \bmod 2^{12}$$

and

$$\tau(8n + 7) \equiv 705\sigma_{11}(8n + 7) \bmod 2^{14}.$$

We will see that these congruences are closely related to the formula for $\delta_{24}(n)$. Here we consider the weight 12 modular form on $\Gamma_0(4)$:

$$q^6\Psi^{24}(q^2) = \frac{\eta^{48}(4\tau)}{\eta^{24}(2\tau)} = \sum_{n=0}^{\infty} \delta_{24}(n)q^{2n+6}.$$

The first few terms of the Fourier expansion are

$$q^6\Psi^{24}(q^2) = q^6 + 24q^8 + 276q^{10} + 2048q^{12} + 11178q^{14} + \cdots.$$

We obtain the following identity:

$$q^6\Psi^{24}(q^2) = \frac{1}{176896}\left[\frac{1}{2048} E(\tau) - \Delta(2\tau) - 2072\Delta(4\tau)\right].$$

$E(\tau)$ is the weight 12 Eisenstein series defined by

$$E(\tau) = \sum_{n=1}^{\infty} \sigma_{11}^{\#}(n)q^n,$$

where $\sigma_{11}^{\#}(n) = \sum_{\substack{d|n \\ n/d \text{ odd}}} d^{11}$. Note that $\sigma_{11}^{\#}(2n + 6) = 2^{11}\sigma_{11}^{\#}(n + 3)$. This identity proves the following formula for $\delta_{24}(n)$.

THEOREM 8. *If $\delta_{24}(n)$ is the number of representations of n as a sum of 24 triangular numbers then*

$$\delta_{24}(n) = \frac{1}{176896}\left[\sigma_{11}^{\#}(n + 3) - \tau(n + 3) - 2072\tau\left(\frac{n + 3}{2}\right)\right].$$

Before proceeding to congruences, we note the connection between $\delta_{24}(n)$ and the Leech lattice. The 24 dimensional Leech lattice is well known for its nice symmetric solution to the sphere packing problem. Let N_m be the number of points on this lattice with norm m. The lattice theta function $\Theta(\tau)$ is defined by

$$\Theta(\tau) = \sum_{m=0}^{\infty} N_m q^m.$$

It turns out that

$$\Theta(\tau) = 1 + 195560q^4 + 16773120q^6 + 398034000q^8 + \cdots$$

is a weight $k = 12$ modular form. Using techniques from modular form theory [3, p. 51], it is known that N_{2m} is given by

$$N_{2m} = \frac{65520}{691}(\sigma_{11}(m) - \tau(m)).$$

We obtain the following corollary to Theorem 8, connecting the representation of $n - 3$ as a sum of 24 triangular numbers with the number of points on the Leech lattice with norm $2n$.

COROLLARY 4. *If $n \geq 3$ odd, then*

$$N_{2n} = N_6\delta_{24}(n - 3). \qquad \square$$

Now we list congruences for $\tau(n)$ that are consequences of this formula. (Note: $176896 = 256 \times 691$). The proofs are left to the reader.

$$\tau(n) + 2072\tau\binom{n}{2} \equiv \sigma_{11}^{\#}(n) \bmod 176896, \tag{2}$$

$$\tau(n) \equiv \sigma_{11}^{\#}(n) \bmod 8, \tag{3}$$

$$\tau(2^k n) \equiv \sigma_{11}(n) \sum_{i=0}^{k} (-2072)^i 2048^{k-i} \bmod 176896, \text{ where } (2, n) = 1, \tag{4}$$

$$\tau(2^k n) \equiv (-24)^k \sigma_{11}(n) \bmod 256, \text{ where } (2, n) = 1. \tag{5}$$

Along these lines we discuss the existence of many formal convolution identities. It will be clear that this phenomenon occurs often when one considers modular forms that are linear combinations of two eigenforms.

First we state and prove a general theorem.

THEOREM 9. *Let $F(n)$, $G(n)$, and $H(n)$ be functions on the non-negative integers such that $F(1) = G(1) = 1$. Suppose α is a non-negative integer where $F(\alpha) \neq G(\alpha)$ and such that*

$$F(\alpha^{n+2}) = F(\alpha)F(\alpha^{n+1}) + H(\alpha)F(\alpha^n)$$

and

$$G(\alpha^{n+2}) = G(\alpha)G(\alpha^{n+1}) + H(\alpha)G(\alpha^n)$$

for all positive integers n. Then

$$\frac{F(\alpha^{n+1}) - G(\alpha^{n+1})}{F(\alpha) - G(\alpha)} = \sum_{k=0}^{n} F(\alpha^n)G(\alpha^{n-k}).$$

Proof. Define two formal power series $f(q)$ and $g(q)$ by

$$f(q) = \sum_{n=0}^{\infty} F(\alpha^n)q^n$$

and

$$g(q) = \sum_{n=0}^{\infty} G(\alpha^n)q^n.$$

By hypothesis, the coefficients of $f(q)$ and $g(q)$ are second-order linear recurrences in n. As a consequence, we obtain the following identities:

$$f(q) = \frac{1}{1 - F(\alpha)q - H(\alpha)q^2} \tag{6}$$

and

$$g(q) = \frac{1}{1 - G(\alpha)q - H(\alpha)q^2}. \tag{7}$$

To prove the theorem it suffices to show that

$$f(q) - g(q) = [F(\alpha) - G(\alpha)]qf(q)g(q).$$

By (6) and (7) we find that

$$\frac{1}{g(q)} - \frac{1}{f(q)} = [F(\alpha) - G(\alpha)]q$$

which proves the theorem.   □

At this time we recall the Dirichlet convolution of two arithmetical functions.

DEFINITION. If $F(n)$ and $G(n)$ are two arithmetical functions, then the Dirichlet convolution $(F * G)(n)$ is an arithmetical function defined by

$$(F * G)(n) = \sum_{d|n} F(d)G\left(\frac{n}{d}\right).$$

For our purposes let $F(n) = \sigma_{11}(n)$, $G(n) = \tau(n)$ and $H(n) = -n^{11}$. Since $E_{12}(\tau)$ and $\Delta(\tau)$ are eigenforms of the Hecke operators T_p, the conditions of the theorem hold for $n = p$ a prime and when the arithmetical functions are chosen to be their Fourier coefficients. We obtain:

$$\frac{\sigma_{11}(p^{n+1}) - \tau(p^{n+1})}{\sigma_{11}(p) - \tau(p)} = \sum_{k=0}^{n} \sigma_{11}(p^k)\tau(p^{n-k}) = (\sigma_{11} * \tau)(p^n). \tag{8}$$

We now apply these ideas to the formula for $\delta_{24}(n)$ given in Theorem 8. Let $n + 3 = p^{k+1}$ where p is an odd prime. By Theorem 8 we find that

$$\delta_{24}(p^{k+1} - 3) = \frac{1}{176896}[\sigma_{11}(p^{k+1}) - \tau(p^{k+1})]. \tag{9}$$

From the convolution identity (8) we obtain the following corollary to Theorem 8.

COROLLARY 5. *If p is an odd prime and k is a positive integer, then*

$$\frac{\delta_{24}(p^{k+1} - 3)}{\delta_{24}(p - 3)} = (\sigma_{11} * \tau)(p^k).$$

Proof. By (8) we obtain

$$\delta_{24}(p^{k+1} - 3) = \frac{\sigma_{11}(p) - \tau(p)}{176896}(\sigma_{11} * \tau)(p^k) = \delta_{24}(p - 3)(\sigma_{11} * \tau)(p^k). \qquad \square$$

5. Representations of integers as sums of figurate numbers

Here we suggest how modular form theory can be used to solve the general problem of calculating the number of representations of integers as sums of figurate numbers. The method requires defining a generating function that is a modular form of weight $\frac{1}{2}$.

The higher figurate numbers are given by the function

$$f_a(n) = \frac{an^2 + (a - 2)n}{2}.$$

Notice that $a = 1$ gives triangular numbers and $a = 2$ gives squares. Here we derive the generating functions for all of these figurate numbers using generalized Dedekind η-functions [21], [22].

DEFINITION. The generalized Dedekind η-function is defined by

$$n_{\delta,g}(\tau) = e^{\pi i P_2(g/\delta)\,\delta\tau} \prod_{\substack{n > 0 \\ n \equiv g \bmod \delta}} (1 - q^n) \prod_{\substack{n > 0 \\ n \equiv -g \bmod \delta}} (1 - q^n),$$

where $P_2(t) = \{t\}^2 - \{t\} + \frac{1}{6}$ is the second Bernoulli polynomial and $\{t\} = t - [t]$ is the fractional part of t.

Theorem 3 of [22] gives a nice criterion for when functions $f(\tau)$ of the form

$$f(\tau) = \prod_{\substack{\delta \mid N \\ 0 \le g < \delta}} \eta_{\delta,g}^{r_{\delta,g}}(\tau),$$

where

$$\tau_{\delta,g} \in \begin{cases} \frac{1}{2}\mathbb{Z} & \text{if } g = 0 \text{ or } g = \delta/2 \\ \mathbb{Z} & \text{otherwise} \end{cases},$$

are modular functions on $\Gamma_1(N)$.

We now establish a relation between generating functions for the figurate numbers and generalized Dedekind η-functions.

THEOREM 10. *If $a \geq 1$, then*

$$q^{(a-2)^2/(8a)} \sum_{n=-\infty}^{\infty} q^{(an^2+(a-2)n)/2} = \frac{\eta(a\tau)\eta_{a,1}(2\tau)}{\eta_{a,1}(\tau)}.$$

Proof. By the Jacobi triple product identity [2]

$$\prod_{n=11}^{\infty} (1-q^{2n})(1+q^{2n-1}z)(1+q^{2n-1}z^{-1}) = \sum_{n=-\infty}^{\infty} q^{n^2}z^n,$$

where $z \neq 0$ and $|q| < 1$. Replacing q by $q^{a/2}$ and z by $q^{(a-2)/2}$, we get

$$\sum_{n=-\infty}^{\infty} q^{(an^2+(a-2)n)/2} = \prod_{n=1}^{\infty} (1-q^{an})(1+q^{(a(2n-1)+a-2)/2})(1+q^{(a(2n-1)-(a-2))/2})$$

$$= \prod_{n=1}^{\infty} (1-q^{an})(1+q^{an-1})(1+q^{an-(a-1)})$$

$$= q^{-a/24}\eta(a\tau) \prod_{n=1}^{\infty} \frac{(1-q^{2an-2})(1-q^{2an-2(a-1)})}{(1-q^{an-1})(1-q^{an-(a-1)})}$$

$$= q^{-a/24}q^{-aP_2(1/a)+(a/2)P_2(1/a)} \frac{\eta(a\tau)\eta_{a,1}(2\tau)}{\eta_{a,1}(\tau)}$$

$$= q^{-(a-2)^2/(8a)} \frac{\eta(a\tau)\eta_{a,1}(2\tau)}{\eta_{a,1}(\tau)}. \qquad \square$$

Notice that the generating functions constructed in the previous theorem are defined as a sum over all integers, in contrast to

$$\Psi(q) = \sum_{n=0}^{\infty} q^{n(n+1)/2}.$$

This difference is minor because

$$\sum_{n=-\infty}^{\infty} q^{n(n+1)/2} = 2 \sum_{n=0}^{\infty} q^{n(n+1)/2}.$$

EXAMPLE. For pentagonal numbers, we choose $a = 3$ in the last theorem. We obtain

$$q^{1/24} \sum_{n=-\infty}^{\infty} q^{(3n^2 + n)/2} = \frac{\eta(3\tau)\eta_{3,1}(2\tau)}{\eta_{3,1}(\tau)} .$$

So, if $\varrho_k(n)$ is the number of representations of n as a sum of k pentagonal numbers then we can calculate $\varrho_k(n)$ using modular forms. Note that $\varrho_k(n)$ will reflect the multiplicity of representations resulting from the possible choices of sign allowed for n in the generating function.

6. The number of lattice points in k-dimensional spheres

Gauss asked how many lattice points are contained in the circle of radius R centered at the origin. It is very difficult to calculate this number as a function of R asymptotically. Although this problem is difficult, it is easy to see that one can calculate the exact number of points in such a circle using $r_2(n)$. One simply must sum $r_2(n)$ for all n up to R.

We generalize this question by asking how many lattice points are contained in a k-dimensional sphere with radius R centered at $(\frac{1}{2}, \frac{1}{2}, \ldots, \frac{1}{2})$? We are interested in this question because it gives a geometric meaning to $\delta_k(n)$.

PROPOSITION 4. *The k-dimensional sphere of radius R centered at $(\frac{1}{2}, \frac{1}{2}, \ldots, \frac{1}{2})$ contains*

$$2^k \sum_{1 \le n \le (R^2/2) - (k/8)} \delta_k(n)$$

lattice points in $\mathbb{Z}^k$.

Proof. Consider concentric spheres centered at $(\frac{1}{2}, \frac{1}{2}, \ldots, \frac{1}{2})$. If a sphere of radius τ contains a lattice point $(x_1, x_2, \ldots, x_k)$, then we have

$$\left(x_1 - \frac{1}{2}\right)^2 + \left(x_2 - \frac{1}{2}\right)^2 + \cdots + \left(x_k - \frac{1}{2}\right)^2 = r^2.$$

This implies that

$$\sum_{i=1}^{k} (x_i^2 - x_i) = r^2 - \frac{k}{4} .$$

Or equivalently,

$$\sum_{i=1}^{k} \frac{(x_i^2 - x_i)}{2} = \frac{r^2}{2} - \frac{k}{8}.$$

We get a representation of $(r^2/2) - (k/8)$ as a sum of k triangular numbers. The number of lattice points on this sphere is $2^k \delta_k((r^2/2) - (k/8))$. Here the scalar 2^k accounts for the fact that x_i and $-x_i - 1$ define the same triangular number. The result now follows.   □

Acknowledgements

The first author would like to thank the Department of Mathematical Sciences at the University of Northern Colorado for their hospitality while he thought about modular forms in July 1993. He would also like to acknowledge the faculty and staff at Woodbury University for the wonderful two years he spent as a faculty member there (1991–1993). He especially thanks Adam and Zelda.

The second author would like to thank Ken and UCLA for their hospitality during the preparation of this paper.

REFERENCES

[1] ANDREWS, G., *Eureka! num* $= \Delta + \Delta + \Delta$. J. Number Theory *23* (1986), 285–293.
[2] ANDREWS, G., *The theory of partitions*. [Encyclopedia of Math., Vol. 2]. Addison-Wesley, Reading, MA, 1976.
[3] CONWAY, J. and SLOANE, N., *Sphere packings, lattices and groups*. Springer-Verlag, 1988.
[4] DELIGNE, P., *Formes modulaires et representations l-adiques*. In Seminaire Bourbaki [Lect. Notes in Math., No. 179], Springer-Verlag, Berlin, 1971.
[5] DELIGNE, P. and SERRE, J.-P., *Formes modulaires de poids 1*. Ann. Scient. Ecole Norm. Sup. (4) *7* (1974).
[6] DICKSON, L., *Theory of numbers, Vol. III*. Chelsea, New York, 1952.
[7] GARVAN, F., KIM, D. and STANTON, D., *Cracks and t-cores*. Invent. Math. *101* (1990), 1–17.
[8] GARVAN, F., *Some congruence properties for partitions that are p-cores*. Proc. London Math. Soc. *66* (1993), 449–478.
[9] GORDON, B. and ROBINS, S., *Lacunarity of Dedekind η-products*. Glasgow Math. J., *37* (1995), 1–14.
[10] GROSSWALD, E., *Representations of integers as sums of squares*. Springer-Verlag, 1985.
[11] HIDA, H., *Elementary theory of L-functions and Eisenstein series*. [London Math. Society Student Text, No. 26]. Cambridge Univ. Press, Cambridge, 1993.
[12] KOBLITZ, N., *Introduction to elliptic curves and modular forms*. Springer-Verlag, Berlin, 1984.
[13] KOLBERG, O., *Congruences for Ramanujan's function τ(n)*. [Årbok Univ. Bergen, Mat.-Natur. Ser. No. 1]. Univ., Bergen, 1962.
[14] LEGENDRE, A., *Traitée des fonctions elliptiques, Vol. 3*, Paris, 1828.
[15] MIYAKE T, *Modular forms*. Springer-Verlag, Berlin, 1989.

[16] ONO, K., *Congruences on the Fourier coefficients of modular forms on* $\Gamma_0(N)$. Contemp. Math, to appear.

[17] ONO, K., *Congruences on the Fourier coefficients of modular forms on* $\Gamma_0(N)$ *with number-theoretic applications*. Ph.D. Thesis, University of California, Los Angeles, 1993.

[18] ONO, K., *On the positivity of the number of t-core partitions*. Acta Arithmetica *66* (1994), 221–228.

[19] RANKIN, R., *Ramanujan's unpublished work on congruences*. [Lect. Notes in Math., No. 601]. Springer Verlag, Berlin, 1976.

[20] RANKIN, R. A., *On the representations of a number as a sum of squares and certain related identities*. Proc. Cambridge Phil. Soc. *41* (1945), 1–11.

[21] ROBINS, S., *Arithmetic properties of modular forms*. Ph.D. Thesis, University of California, Los Angeles, 1991.

[22] ROBINS, S., *Generalized Dedekind η-product*. To appear in Contemp. Math.

[23] SCHOENEBERG, B., *Elliptic modular functions — an introduction*. Springer-Verlag, Berlin, 1970.

[24] SERRE, J. P., *Sur la lacunarite' des puissances de η*. Glasgow Math. J. *27* (1985), 203–221.

[25] SERRE, J. P., *Quelques applications du theorme de densite de Chebotarev*. [Publ. Math. I.H.E.S., No. 54]. Inst. Hautes Etudes Sci. Pub., I.H.E.S., Paris, 1981.

[26] SERRE, J. P., *Congruences et formes modulaires (d'apres H.P.F. Swinnerton-Dyer)*. In Seminaire Bourbaki, 24e anneé (1971/1972), Exp. No. 416. [Lect. Notes in Math., No. 317]. Springer Verlag, Berlin, 1973, pp. 319–338.

[27] SERRE, J.-P. and STARK, H., *Modular forms of weight* $\frac{1}{2}$. In modular functions of one variable, Vol. VI. [Lect. Notes in Math., No. 627]. Springer Verlag, Berlin, 1971, pp. 27–67.

[28] SHIMURA, G., *Introduction to the arithmetic theory of automorphic functions*. [Publ. Math. Soc. of Japan, No. 11], Iwanami Shoten, Tokyo, 1971.

[29] SWINNERTON-DYER, H. P. F., *On l-adic representations and congruences for coefficients of modular forms*. [Lect. Notes in Math., No. 350]. Springer Verlag, Berlin, 1973.

[30] SWINNERTON-DYER, H. P. F., *On l-adic representations and congruences for coefficients of modular forms II*. [Lect. Notes in Math., No. 601]. Springer Verlag, Berlin, 1976.

Department of Mathematics,
The University of Georgia,
Athens, Georgia 30602,
U.S.A.
E-mail address: ono@joe.math.uga.edu

Department of Mathematical Sciences,
University of Northern Colorado,
Greeley, Colorado 80639,
U.S.A.
E-mail address: srobins@dijkstra.univnorthco.edu

Department of Mathematics,
University of Colorado,
Boulder, Colorado 80309,
U.S.A.
E-mail address: pwahl@euclid.colorado.edu

Aequationes Mathematicae **50** (1995) 95–134
University of Waterloo

0001–9054/95/020095–40 $1.50 + 0.20/0
© 1995 Birkhäuser Verlag, Basel

Chromatic sums revisited

W. T. Tutte

Summary. This paper discusses some equations arising in the author's work on "chromatic sums". The main results were presented in a series of papers extending from 1973 to 1982. The object here is to give a unified and simplified account of the elimination of unwanted variables from the initial equation, leading to the final identification of the desired chromatic sums as the coefficients in a power series satisfying a certain differential equation.

1. Introduction

Let us begin by stating two equations that are of major importance in what follows.

$$xg = x^3 y \lambda(\lambda - 1) + \lambda^{-1} yzgq + yz(g - x^2 L) - x^2 y^2 z \, \Delta g. \tag{α}$$

$$H''\{2t + 5H - 3tH'\} = 48t \qquad (t = z^2). \tag{ω}$$

In equation (α) the symbol g stands for a power series in four variables x, y, z and λ, the indices being integral and non-negative. Substituting 1 for y in g we obtain the power series q, and L is defined as the coefficient of x^2 in g. Thus L is a power series in y, z and λ. The symbol Δ represents an operator acting on functions of y. Applied to such a function $F(y)$ it gives the quotient

$$(F(y) - F(1))/(y - 1).$$

AMS (1991) subject classification: 05C15.

Manuscript received August 9, 1993 and, in final form, March 9, 1994.

Thus we have

$$(y - 1)\, \Delta g = g - q. \tag{1}$$

In theory, equation (α) can be solved by a recursive procedure that calculates in order the coefficients of successive powers of z in g. Thus by picking out coefficients of z^0 in (α) we find that the coefficient of z^0 in g must be $\lambda(\lambda - 1)x^2 y$. So the corresponding coefficients in q and L are $\lambda(\lambda - 1)x^2$ and $\lambda(\lambda - 1)y$ respectively. Now by picking out the coefficients of z^1 in (α) we can find the coefficient of z^1 in g, and therefore in q and L, and so on.

In this paper we try to show that there is more to the theory of g and its specializations than this possibility of recursive calculation. However we infer from the above argument that equation (α) has a unique solution for the power series g, and that in this solution the coefficient of each power of z is a polynomial in the other three variables. The latter property ensures that the substitution of 1 for y is meaningful.

The author was interested in a power series h defined as the result of substitution 1 for y in L or equivalently as the coefficient of x^2 in q. This power series is a function of only two variables z and λ. The author strove to deduce from (α) an equation for h involving only these two variables, and after nine years he succeeded. In equation (ω) we have the special case $\lambda = 4$ of the final equation, that case having an unusually simple form and being of exceptional interest to students of chromatic polynomials and chromatic sums. In (ω) H is to be understood as $t^2 h$, and a prime denotes differentiation by t.

Following a suggestion of a referee we conclude this section with a brief account of the contents of the following ones "to serve as a chart". We note in each case what new terminology and important symbols are introduced.

Section 2 defines g, q, L and h as enumerating functions of combinatorial objects, noting that as such they have been shown to satisfy equation (α). In each case the main variable is z and the coefficients of its powers are polynomials in the other variables x, y and λ.

Section 3 is concerned with some changes of variables that are thought convenient. Thus x is replaced by a new variable s satisfying $x = zs$, and z is replaced by $t = z^2$. Equation (α) and the enumerating functions are stated in terms of the new variables. New symbols μ and v, representing simple functions of λ, are introduced. A change of variable from y to a simple function V of y is foreshadowed, though it will not come into effect until Section 10.

Section 4 tells what kind of power series are to be used in the paper. The main variable t is to occur only to non-negative powers. A power series is to be taken either from the set $Y1$ in which the coefficients of the powers of t are polynomials in the other variables with real numbers as coefficients, or from the set $Y2$ in which the coefficients of the powers of t are quotients of such polynomials. A notation with such symbols as $X1(y, \lambda)$ and $Y2(s, \lambda)$ is set out. The operations of division, taking square roots and substitution for such power series are discussed. The author resolves not to divide by a power series with a zero constant terms.

In Section 5 we eliminate from (α) the 4-variable function g, at the expense of introducing two new 3-variable functions y_1 and y_2 (of s, t and λ). We end up with four equations for L, q, y_1 and y_2.

In Section 6 the enumerating series q is eliminated and the four equations are replaced by three.

Section 7 is in the nature of a check. It verifies that the three equations do indeed determine L, y_1 and y_2 uniquely.

In Section 8 we define an "invariant", that is, a function of y that takes the same value when either y_1 or y_2 is substituted for y. Power series not involving y can be counted as trivial invariants. A function J is defined, closely related to L, and it is pointed out that J is an invariant. So, of course, is any polynomial in J with coefficients depending only on t and λ. In later sections J is to be replaced by $K = J + v$. The section goes on to consider hypothetical invariants of a special kind, "members of W_k", showing how each such invariant can be expressed as a polynomial in J (or K). Examples of such special invariants will occur in Section 10. They involve trigonometric functions.

Section 9 lists some trigonometrical identities needed in our theory of special invariants. The symbol Q_n, representing in effect a "Beraha polynomial" is introduced. The symbol α is used for an arbitrary non-zero real angle.

In Section 10 we replace y by V. Substituting y_1 or y_2 for y in V, we obtain V_1 and V_2, respectively. Similarly from K we get K_1 and K_2. But the last two are equal since K is an invariant. Their common value is expressed in terms of V_1 and V_2. At this stage we specialize α to the value $2\pi/N$, where $N = 2M$ is a positive even number. We also assign the value $2 \cos \alpha$, that is, $2 \cos(2\pi/N)$ to μ, thus substituting the number $2 + 2 \cos(2\pi/N)$ for the variable λ. This substitution for λ changes K, K_1 and K_2 into $K(N)$, $K_1(N)$ and $K_2(N)$ respectively.

We go on to define $E(r)$, for any positive integer r, as a function of $K(N)$, V and the Q_n, the last defined in terms of the special α. Substituting V_i for V in $E(r)$ we get $E_i(r)$ ($i = 1, 2$). We find for $E_i(r)$ a simple expression in terms of V_1, V_2 and the Q_n. Finally we discover that the product of $E_i(r)$ from $r = 1$ to $r = M - 1$, multiplied by $v^2 - V^2$, is an invariant $I(N)$. It even "belongs to W_{M-2}" and so, by the theorem of Section 8, is a polynomial in $K(N)$ of degree $M - 2$. The coefficient g_i of $(K(N))^i$ is a power series in t with numerical coefficients. The author apologises for a second use of the symbol g.

In Section 11 we apply the trigonometrical identities of Section 9 to $E(r)$ and so arrive at a trigonometrical formula, equation (63), for $I(N)$. We work in a new domain $Y3(V, \lambda)$ of power series.

In Section 12 we eliminate $I(N)$ from its expressions found in Sections 10 and 11 to get an "exploitable equation" and hence deduce expressions for g_{M-2} and g_{M-3}. The first coefficient we get explicitly as a numerical multiple of t; the second is a linear form in the enumerating function h. We find also that each g_i has a zero constant term as a power series in t.

In Section 13 we discuss the substitution for V of a power series f in t with real numbers as coefficients. This converts $K(N)$ into $K(N, f)$. It is shown that, if the constant term b of f is fixed as a number not 0, v, $v - 1$ or $2v/\mu$, then f can be extended uniquely as a solution of the functional equation $f^2 = cK(N, f)$, where $c = b^2/(b^2 - \mu vb + v^2)$. The four excluded numbers are rejected for various reasons, some because of the author's resolve in Section 4.

In the application of this theorem c is taken to be the square of $\sin \theta/\sin(2\pi/N)$, and θ is subsequently restricted to be of the form $\pi r/N$, where r is an integer. We thereby arrange for an important term of the "exploitable equation" to be reduced to triviality when f is substituted for V. We are left with an algebraic equation in a parameter $w_S = (4 \sin^2(\pi r/N))/f^2$ where S is a positive integer depending on r. For M at least δ, that is N at least 16, we distinguish $M - 2$ such parameters, with S ranging from 1 to $M - 2$. The restriction on M is to avoid forbidden values of the corresponding b. (We add $\mu v/2$ to the list of forbidden values for a reason made clear in the next Section).

Section 14 is concerned with the algebraic equations satisfied by each parameter w_S. First we define p_j as a linear form in g_{M-j}, with numerical coefficients depending on j and M. We then discover that the sum of $p_j w_S^j$ from $j = 0$ to $j = M$ is $(-1)^S$. Then we substitute f for V in the V-derivative of the "exploitable equation" and find that the sum of $j p_j w_S^{j-1}$ over the range of j is zero. These are our two "parametric equations" for w_S. (The derivation of the second fails when $b = \mu v/2$), which is why we eschew that value of b.)

In Section 15 we introduce a new variable w and define U as the sum of $p_j w^j$ from $j = 0$ to $j = M$. We use the symbol D_x to denote differentiation with respect to x. We deduce from the parametric equations that the $M - 2$ parameters are zeros of $D_w U$, double zeros of $U^2 - 1$ and zeros of $D_t U$. These observations lead to the partial differential equations (93) and (94). The first involves initially unknown power series α and β in t with numerical coefficients depending on N. The second involves a polynomial Y in w whose coefficients are similar initially unknown power series A, B, C and D. A routine manipulation of the partial differential equations gives the partial differential equation (106) for Y.

Using the definition of p_j and the evaluation of g_{M-2} and g_{M-3} in Section 12 we determine α and β, the first as $-\lambda/2$ and the second as a linear function of h.

In Section 16 we rewrite the equation for Y as a polynomial identity in w. Then, by equating coefficients, we get five simultaneous differential equations for A, B, C and D. Their coefficients involve α and β, that is depend on h. Eliminating A, B, C and D from these we get a differential equation for an integral γ of β, which we then rewrite as equation (127) for H, that is t^2h. Its coefficients are given as simple polynomials in λ. But we need to remember that it is proved only for some, though infinitely many, special values of λ.

In Section 17 we observe that (127) does have a unique solution H for general λ such that the coefficients of t^0 and t^1 are zero. In that H, the coefficients of the powers of t are polynomials in λ. By our previous work, this H agrees with our enumerative H for infinitely many values of λ. Hence by their common property of polynomial coefficients the two H's must be identical: equation (127) must hold for general λ.

Section 18 comments on some special values of λ. In particular equation (ω) is derived by substituting $\lambda = 4$ in (127).

2. Chromatic sums

Since the early 1960's some combinatorialists have studied the problem of enumerating rooted planar maps of various kinds. In "rooting" a planar map we label one of its edges as the "root-edge" and specify positive directions along and across this edge. For maps not encumbered with loops or isthmuses we may instead choose a "root-vertex", a "root-edge" and a "root-face", mutually incident. The positive direction along the root-edge is then from the root-vertex to the other end. The positive direction across the root-edge is from the root-face to the face on the other side. Conventionally the root-face is taken to be the outer one.

We are concerned in this paper with the planar maps called "triangulations". A triangulation is a non-separable planar map in which each face is a triangle. We also define a "near-triangulation" as a non-separable planar map which has at most one non-triangular face. Thus every triangulation counts as a near-triangulation. For algebraic convenience we recognize the "link-map" as a near-triangulation. This trivial figure has just one edge, two incident vertices and a single face, counted as a digon. All rootings of the link-map are symmetrically equivalent. We therefore say that there is just one rooted link-map. Every other near-triangulation has at least one triangular face, and therefore at least two of them. We impose the restriction that, when a near-triangulation is rooted, the non-triangular face, if there is one, must be the root-face.

Usually a researcher on triangulations drags near-triangulations into his theory only reluctantly, seeing no way of constructing an interesting functional equation without them.

Two questions arise. How many non-isomorphic rooted planar maps are there with $2n$ triangles? How many non-isomorphic rooted near-triangulations are there with p triangular faces and a root-face of valency m? These two questions are answered in [2], using generating series.

It seemed natural to generalize the problem as follows. How many λ-coloured rooted planar triangulations are there with $2n$ triangles? We define a λ-colouring of a map M, rooted or non-rooted, with λ a positive integer, as a colouring of the vertices of M from λ given colours so that no two of the same colour are joined by an edge. A triangulation or near-triangulation is said to be "λ-coloured" when it is given a λ-colouring.

The number of λ-colourings of M is denoted by $P(M, \lambda)$. It is found to be a polynomial in λ whose degree is the number of vertices of M. It is therefore called the "chromatic polynomial" of M. Since it is a polynomial it takes definite values even for real or complex values of λ that are not positive integers. But in the positive integral case the number of λ-coloured rooted maps derived from a given rooted map M is $P(M, \lambda)$. Evidently the chromatic polynomial of the link-map is $\lambda(\lambda - 1)$.

For any given rooted near-triangulation M we write $p(M)$ for the number of non-root triangular faces, $m(M)$ for the valency of the root-face and $n(M)$ for the valency of the root-vertex. Then we define a power series g in x, y, z and λ as follows:

$$g = \sum_M x^{m(M)} y^{n(M)} z^{p(M)} P(M, \lambda). \tag{2}$$

We note that the typical coefficient in the above sum, λ being a positive integer, is the number of rooted λ-coloured near-triangulations with specified values of $m(M)$, $n(M)$ and $p(M)$. But we can also regard λ as a variable for which any real or complex number may be substituted. In general we refer to the typical coefficient in (2) as a "chromatic sum" taken over all rooted near-triangulations with those specified values of $m(M)$, $n(M)$ and $p(M)$.

Having established g, we define q and L from it as in Section 1. It can then be shown by simple graph-theoretical arguments that the three power series satisfy equation (α). There is a proof in [3], pages 427–432. The coefficients in these power series too are chromatic sums, taken over appropriate classes of maps. In [3] the symbol of the series L is lower case. But that letter is too easily confused with the numeral 1.

Now L is the coefficient of x^2 in g. It is a sum over those rooted planar maps whose root-face is a digon. Among these we notice the link-map, in which the two edges of the digon coincide, and which makes the contribution $\lambda(\lambda - 1)y$ to L. In every other contributing rooted map M the two edges of the digon are distinct. When we delete the one that is not the root-edge, we convert M into a rooted triangulation T satisfying

$$p(T) = p(M), n(T) = n(M) - 1. \tag{3}$$

This operation being reversible we can write L as, barring the contribution of the link-map, a sum over the rooted planar triangulations T:

$$L = y\left\{\lambda(\lambda - 1) + \sum_T y^{n(T)} z^{p(T)} P(T, \lambda)\right\}. \tag{4}$$

Substituting 1 for y in this equation we find

$$h = \lambda(\lambda - 1) + \sum_T z^{p(T)} P(T, \lambda). \tag{5}$$

So the typical coefficient in h is the chromatic sum over all rooted planar triangulations with a given number of triangles. This is the power series for which the author looked throughout his succession of papers on chromatic sums. The rest is but scaffolding. We write $H = t^2 h$ as in Section 1.

Perhaps we now have a sufficient clarification of the author's motivation in studying equation (α). That equation is not a random doodle; it arises naturally out of an interesting combinatorial problem.

3. Changes of variables

Workers on chromatic polynomials use sometimes λ, sometimes $(\lambda - 1)$ and sometimes $(\lambda - 2)$ as their basic variable, whichever seems to give the neatest formulae. We shall often adopt the third suggestion, writing

$$\mu = \lambda - 2. \tag{6}$$

As another convenient abbreviation we shall write

$$v = 1/(2 - \mu) = 1/(4 - \lambda). \tag{7}$$

The original variables may seem convenient when the emphasis is on the chromatic interpretation of the coefficients in g, q and L. But when we embark on an algebraic analysis of equation (α) some minor changes seem desirable. First we replace x by a new variable s related to it as follows.

$$x = zs. \tag{8}$$

Let us make this substitution in equation (α) and then divide its terms by their common factor z. We obtain a modified equation in which z now appears explicitly only to the second power. We can infer from this that in each of the power series obtained by the recursive solution the variable z now occurs only to even powers. We can get that result directly; by a simple graph-theoretical argument the number $p(M) + m(M)$ must be even. Let us therefore replace z in our formulae by the new variable

$$t = z^2, \tag{9}$$

the variable of equation (ω).

In terms of the new variables we have

$$g = \sum_M s^{m(M)} y^{n(M)} t^{r(M)} P(M, \lambda), \tag{10}$$

where

$$2r(M) = p(M) + m(M). \tag{11}$$

As before, we obtain the formula for q by substituting 1 for y in the formula for g. But now L is the coefficient of s^2 in g, divided by t. However, h is still the result of substituting 1 for y in L.

In every rooted planar near-triangulation M the number $m(M)$ is at least 2, and therefore the number $r(M)$ is at least 1, by (11). It is exactly 1 only for the link-map. We can thus state the initial terms of our power series, in their new guise, as follows.

THEOREM 3.1. *The coefficient of t^0 in g and in q is zero. In L it is $\lambda(\lambda - 1)y$ and in h it is $\lambda(\lambda - 1)$.*

In terms of the new variables, equation (α) can be written as follows:

$$sg = s^3 ty\lambda(\lambda - 1) + \lambda^{-1} ygq + y(g - s^2 tL) - s^2 ty^2 \Delta g. \tag{12}$$

At a later stage of this work a more drastic change will be almost forced upon us. It will replace y by a new variable V related to y as follows.

$$y^{-1} = V + 1 - v. \tag{13}$$

4. Domains of coefficients

This section is to explain the nature of the formal power series to be used in this paper. They are power series in the variable t, the indices of t being integral and non-negative. Let $X1$ denote the set of all polynomials in the other variables, with real numbers as coefficients. Let $X2$ be the set of all quotients P/Q, where P and Q belong to $X1$. We allow Q to be unity, so we can regard $X1$ as a subset of $X2$. In general, let the variables other than t be $u_1, u_2, \ldots, u_k$. To show them explicitly, we extend the symbol Xj, where $j = 1$ or 2, as follows: $Xj(u_1, u_2, \ldots, u_k)$. A member of $Xj(u_1, \ldots, u_k)$ need not involve all of these variables, but must use no others. It may even involve none of them, reducing to a real number.

We define $Y1$ and $Y2$ as the sets of those power series in t in which the coefficients of the various powers of t belong to all $X1$ or $X2$ respectively. We extend these symbols like those of $X1$ and $X2$ when we wish to indicate the variables other than t. Thus the power series g, q, L and h defined in Section 2 and subjected to a change of variables in Section 3 are members of $Y1(s, y, \lambda)$. We can say further that q belongs to $Y1(s, \lambda)$, L to $Y1(y, \lambda)$ and h to $Y1(\lambda)$. After the change from y to V we shall write of $X2(s, V)$.

We note that $X1$ and $X2$ are closed under addition and multiplication. Hence $Y1$ and $Y2$ are closed under these operations, including multiplication by a member of $X1$ or $X2$ respectively.

In Section 1 we remarked that for each of the power series under consideration the coefficient of each power of t was a polynomial in the other variables, including λ, with integral coefficients. It is true that the reciprocal of λ occurs in formulae (α) and (12) but it always cancels. This is because the chromatic polynomial of every rooted planar map M divides by λ. In other words one cannot colour M with no colours at all. The power series still have this polynomial property when they are stated in terms of s, y, t and λ.

We shall sometimes want to take reciprocals and square roots of formal power series. Consider a member $A = 1 + tB$ of $Y1$, where B is another member of $Y1$. Then A has a reciprocal in $Y1$. To get it we expand the form $(1 + tB)^{-1}$ by the binomial theorem as a power series in tB. We multiply out each term $(-1)^r (tB)^r$ of that expansion as a power series in t with zero coefficients for all powers of t lower than the rth, and add the results together. Here we have a formal infinite sum, but it reduces to a finite one for each individual power of t. Similarly we find that A has a square root in $Y1$. The argument is equally valid when "$Y1$" is replaced by "$Y2$" throughout.

Now consider a member A of $Y2$ of the form $A = (a + tB)$, where a is a non-zero member of $X2$ and B is in $Y2$. Then A has a reciprocal in $Y2$ since it can be written as $a(1 + ta^{-1}B)$. If $a = b^2$, where b is a non-zero member of $X2$, then A has a square root in $Y2$ since it can be written as $b^2(1 + tb^{-2}B)$. We shall extract such a square root in the next section.

We must also consider the operation of substituting, in a member A of $Y2$, a second member C of $Y2$, for some variable u other than t. That is, we replace u by C whereever it occurs in A. In general the result is a well-defined member D of $Y2$, but there are exceptions to this rule. Consider the coefficient P/Q of t^k in A. There is no difficulty in substituting C for u in the polynomial P; only multiplication and finite addition are involved. Write $P(C)$ for the resulting member of $Y2$. Similarly we substitute C in Q to get a member $Q(C)$ of $Y2$. It may happen that $Q(C)$ is of the above form $(a + tB)$, with a non-zero. Then $Q(C)$ has a reciprocal in $Y2$. We have only to multiply this reciprocal by $t^k P(C)$ to get the contribution to D of the term of A in t^k. We finally arrive at D by summing over k, this operation reducing to a finite sum for each individual power of t.

Our substitution becomes dubious when $Q(c)$ is not of the form $(a + tB)$ with a non-zero, that is when $Q(c)$ is divisible by t. For one thing there is a danger of ending with a power series involving negative powers of t and so not belonging to $Y2$. We shall deal with this case simply by avoiding it.

We might consider generalizing $Y2$ by allowing algebraic functions as coefficients. We shall make one rather trivial venture in this direction later in the paper.

5. Elimination of g

We begin our analysis of (12) by using (1) to express the left side in terms of Δg. After a rearrangement we have

$$\{(y - 1)(s - \lambda^{-1}yq - y) + s^2ty^2\}\, \Delta g = \lambda(\lambda - 1)s^3ty - s^2tyL - q(s - \lambda^{-1}yq - y).$$

$$(14)$$

The coefficient of Δg on the left, a quadratic form in y, can be written as

$$F = (s^2t - \lambda^{-1}q - 1)y^2 + (s + \lambda^{-1}q + 1)y - s. \tag{15}$$

Its discriminant is

$$D = (s + \lambda^{-1}q + 1)^2 + 4s(s^2t - \lambda^{-1}q - 1). \tag{16}$$

Referring to 3.1 we see that this can be written as

$$D = (s - 1)^2 + tC, \tag{17}$$

where C is in $Y2$, but does not involve y, and each of its coefficients can be written as a polynomial in s and λ. We infer that D has a square root d in $X2$ wherein each coefficient of a power of t can be written as a polynomial in s and λ, divided by a power of $s - 1$. The coefficient of t^0 in d is $s - 1$.

We deduce that F has two roots in $Y2$, two power series not involving y, whose substitution for y in F makes that quadratic form vanish. Let us denote them by y_1 and y_2. By the usual formula for the solution of a quadratic we can put

$$2(s^2t + \lambda^{-1}q - 1)y_1 = -(s + \lambda^{-1}q + 1) + d, \tag{18}$$

$$2(s^2t - \lambda^{-1}q - 1)y_2 = -(s + \lambda^{-1}q + 1) - d. \tag{19}$$

Using 3.1 we find that the coefficients of t^0 in y_1 and y_2 are 1 and s respectively. Next we write the equations for the sum and product of the roots.

$$(s^2t - \lambda^{-1}q - 1)(y_1 + y_2) + (s + \lambda^{-1}q + 1) = 0, \tag{20}$$

$$(s^2t - \lambda^{-1}q - 1)y_1y_2 + s = 0. \tag{21}$$

Let us now consider the effect of substituting y_j for y in equation (14). ($j = 1, 2$). The left side vanishes. Let us expand the symbol L as $L(y, t, \lambda)$, showing the three variables involved in that power series. The substituted L we write as $L(y_j, t, \lambda)$. After the substitution we have

$$0 = \lambda(\lambda - 1)s^3ty_j - s^2ty_jL(y_j, t, \lambda) - q(s - \lambda^{-1}y_jq - y_j) \qquad (j = 1, 2). \tag{22}$$

We have now got rid of the 4-variable power series g. Instead we now have four equations, in formulae (20), (21) and (22), for the four functions L, q, y_1 and y_2, each of which involves three variables only. We postpone any discussion of whether this change is for the better until we have made the simplification of the next section.

6. Elimination of q

We can think of the equations (20) and (21) as simultaneous equations for the "unknowns" s and q. Solving them as such, we find

$$sty_1y_2 + (1 - y_1)(1 - y_2) = 0, \tag{23}$$

$$y_1^2y_2^2t(\lambda^{-1}q + 1) = (y_1 - 1)(y_2 - 1)(y_1y_2 - y_1 - y_2). \tag{24}$$

We can use these results to eliminate s and q from (22). We then get the two following equations:

$$(1 - y_1)y_1y_2^2tL(y_1, t, \lambda) = \lambda(1 - y_1(1 - y_2)(y_1 - \mu y_2 + \mu y_1y_2) + \lambda ty_1^2y_2^2, \tag{25}$$

$$(1 - y_2)y_1^2y_2tL(y_2, t, \lambda) = \lambda(1 - y_1)(1 - y_2)(y_2 - \mu y_1 + \mu y_1y_2) + \lambda ty_1^2y_2^2. \tag{26}$$

In the deduction of these equations there is a cancellation of $(1 - y_1)$ and another of $(1 - y_2)$. These cancellations are legitimate since neither power series is zero, by (23).

The twin equations (25) and (26) are found in [7] by a more laborious method. They are the title-equations of [8]. It is said in those papers that the two equations suffice to determine the function L. But this is doubtful unless (23) can also be used, bringing the number of equations for the three unknown functions y_1, y_2 and L up to 3. The next section is devoted to a demonstration that (23), (25) and (26) do indeed determine those power series uniquely.

7. Another recursive procedure for L

We now wish to verify that equations (23), (25) and (26) can be solved for L, y_1 and y_2, as power series in $Y1(y, t, \lambda)$, $Y2(s, t, \lambda)$ and $Y2(s, t, \lambda)$ respectively. We first apply (23) to (25) and (26) to get the two latter equations into the following shorter forms.

$$(1 - y_1)y_2L(y_1, t, \lambda) = -\lambda s(y_1 - \mu y_2 + \mu y_1y_2) + \lambda y_1y_2, \tag{27}$$

$$(1 - y_2)y_1L(y_2, t, \lambda) = -\lambda s(y_2 - \mu y_1 + \mu y_1y_2) + \lambda y_1y_2. \tag{28}$$

For the coefficient of t^j in L, y_1 or y_2 we write $L[j]$, $y_1(j)$ and $y_2(j)$ respectively. For the coefficient of y^k in $L[j]$ we write $L[j, k]$.

Picking out constant terms in (23) we find that either $y_1(0)$ or $y_2(0)$ is unity, and we can adjust the notation so that

$$y_1(0) = 1. \tag{29}$$

From this result the constant term on the left of (27) is zero. So, by picking out constant terms in (27) and using (29), we find that

$$y_2(0) = s. \tag{30}$$

These two formulae agree with the results of Section 5.

Now we pick out the constant terms in (28), finding from (29) and (30) that

$$(1 - s) \sum_k L[0, k]s^k = -\lambda s(s - \mu + \mu s) + \lambda s = \lambda s(1 - s) + \lambda \mu s(1 - s), \tag{31}$$

$$L[0] = \lambda(\lambda - 1)y. \tag{32}$$

We recognize here the contribution of the link-map to the function L of Section 2.

We can repeat the procedure for coefficients of t^1, then of t^2, and so on. To justify this statement suppose we have found the coefficients of the three functions up to those of t^r, and that we now wish to determine those of t^{r+1}. First we equate coefficients of t^{r+1} in (23). Since $(1 - y_1)$ has a zero constant term we find that

$$y_1(r + 1) \cdot (1 - s) + A = 0,$$

where A is calculable as a rational function of s and λ from coefficients we already know. So the above equation determines $y_1(r + 1)$.

For example if $r = 0$ we have $A = s^2$, by (23). Hence

$$y_1(1) = s^2/(s - 1). \tag{33}$$

Next we equate the coefficients of t^{r+1} in (27). On the left, since $(1 - y_1)$ has a zero constant term, we get a member of $X2\,(s, \lambda)$, calculable from the coefficients we already know, including $y_1(r + 1)$. On the right we get another such member of $X2$ (s, λ), plus $\lambda y_2(r + 1)$. So (27) now determines $y_2(r + 1)$.

Finally we equate coefficients of $t^{(r+1)}$ in (28). On the left, apart from a calculable member of $X2(s, \lambda)$ we have

$$(1 - s) \sum_k L[r + 1, k]s^k,$$

and on the right we have another calculable member of $X2(s, \lambda)$. We end up with the equation

$$(1 - s) \sum_k L[r + 1, k]s^k = A,$$

where A is a calculable member of $X2(s, \lambda)$.

Now L belongs to $Y1(y, t, \lambda)$, so in any acceptable solution the sum on the left of (34) is finite. That is, $L[r + 1, k]$ is non-zero for only a finite number of values of k. If A is a polynomial in s that divides by $(1 - s)$ such a solution can be found by first dividing the equation by $(1 - s)$ and then equating coefficients of like powers of s. The coefficient $L[k]$ is then uniquely determined. If A is not such a polynomial then no acceptable solution of (34) exists and the triad of equations (23), (27) and (28) cannot be solved under the stated restrictions. That possibility is ruled out by our knowledge that such a solution of the triad does exist, L being the enumerative function of Section 2 and y_1 and y_2 being given in terms of the enumerative function q by equations (18) and (19).

We infer that the recursion succeeds and that the resulting power series L, y_1 and y_2 are those of the preceding sections.

The triad of equations (23), (27) and (28) can now serve as an alternative definition of L, y_1 and y_2, a definition that is independent of q. We can now claim to have made some progress in our analysis of equation (α). That equation involves three functions, one of four variables and the other two of three each. The new equations still involve three power series but each of them is of three variables only. It is not a valid criticism that we now have three equations instead of one, for (α) had to be buttressed by two rules giving q and L in terms of g.

The author sees no short and direct way from our triad to equation (ω). In papers [7] and [8] we proceed through a theory of "invariants" to another exploitable equation. We explain that theory in the next section.

8. Invariants

In [7] we were introduced to the function

$$J = \lambda^{-1} tL(y, t, \lambda) - yt(1 - y)^{-1} + y^{-2} - y^{-1}, \tag{35}$$

a member of $Y2(y, \lambda)$. In it all the coefficients of powers of t are polynomials in y and λ, except for those of t^0 and t^1.

Members J_1 and J_2 of $Y2(s, \lambda)$ are now defined as the results of substituting y_1 and y_2 respectively for y in J. Our hasty resolve (in Section 4) not to divide by power series with a zero constant term causes some embarrassment when we try to substitute y_1 for y in the term $yt(1-y)^{-1}$. We had better state explicitly that $1 - y_1 = tA$, where A is in $Y2(s, \lambda)$, and the term to be evaluated is to be interpreted as yA^{-1}. We know that A has a non-zero constant term, by (33).

We can substitute y_1 or y_2 for y in (35) and then eliminate $L(y_j, t, \lambda)$ from the formula by using (25) or (26). In [7] it is found that this procedure gives the same result for both substitutions. We have

$$J_1 = J_2 = y_1^{-2} - \mu y_1^{-1} y_2^{-1} + y_2^{-2} + (\mu - 1)\{y_1^{-1} + y_2^{-1}\} - \mu,$$

the expression on the right being symmetrical in y_1 and y_2.

For the purposes of this paper we define an "invariant" as a member of $Y2(y, \lambda)$ such that the substitution of either y_1 or y_2 for y gives the same member of $Y2(s, \lambda)$. We note that any sum, difference or product of invariants is an invariant, and that the above function J is an invariant. Also each member of $Y2(\lambda)$, being independent of y, is trivially an invariant. So any polynomial in J with coefficients in $Y2(\lambda)$ is an invariant.

After the change from y to V we shall find it convenient to use the invariant

$$K = J + v \tag{36}$$

of $Y2(y, \lambda)$ instead of J. When we do this we must exclude the case $\mu = 2$, that is $\lambda = 4$, from consideration, by (7).

Let us write $Y2A(y, \lambda)$ for the set of those power series in t wherein each coefficient is a finite sum of terms each of which is an integral power of y multiplied by a member of $X2(\lambda)$. The index of that integral power may be positive, negative or zero. Thus $Y2A(y, \lambda)$ contains $Y1(y, \lambda)$ and is a subset of $Y2(y, \lambda)$.

THEOREM 8.1. *Let a member F of $Y2A(y, \lambda)$ be an invariant. Then F is a trivial invariant, that is independent of y.*

Proof. If possible, let F be an invariant of $Y2A(y, \lambda)$ for which the theorem fails.

By subtracting a trivial invariant we can reduce to the case in which the coefficient of y^0 is zero at every power of t.

Not all the coefficients of powers of t in F are zero. Let the first non-zero one be the coefficient C_r of t^r. Substituting y_1 for y in F we get a power series in t in which the coefficient of t^r is a function of λ only. But substituting y_2 instead, gives a power series in which the coefficient of t^r is the same function of s and λ that C_r

is of y and λ. So the two substitutions give different members of $Y2(s, \lambda)$, and accordingly F is not an invariant. This contradiction establishes the Theorem.

For the purposes of the next theorem we define W_k, where k is a non-negative integer, as the set of all members of $Y2(y, \lambda)$ of the form

$$\frac{Gt^k}{(1-y)^k},$$

where G is a member of $Y2A(y, \lambda)$.

THEOREM 8.2. *Let F be an invariant belonging to W_k. Then there are members $a_0, a_1, a_2, \ldots, a_k$ of $Y2(\lambda)$ such that*

$$F = \sum_{j=0}^{k} a_j J^j.$$

Proof. The theorem is true for $k = 0$, by 8.1. Assume as an inductive hypothesis that it is true for all values of k less than some positive integer d, and consider the case $k = d$. Some elementary algebraic manipulation puts F into the following form.

$$F = A + \frac{a(k)t^k}{(1-y)^k},$$

where A is in W_{k-1} and $a(k)$ in $Y2(\lambda)$. An application of (35) now shows that

$$F = B + a(k)J^k.$$

where B is in W_{k-1}. Since F and $a(k)J^k$ are invariants so also must be B. The Theorem follows by induction.

We shall have to consider some invariants in $Y2(y)$. We can derive one from J or K by substituting a fixed number c for the variable λ. We now define a "special invariant" as an invariant of $Y2(y)$ that is derived from a member of $Y2(y, \lambda)$, not necessarily itself an invariant, by such a substitution. We shall discover some special invariants in Section 10.

The special invariants of Section 10 involve trigonometrical functions. Accordingly we interpolate Section 9 which takes note of the relevant trigonometrical formulae.

9. Some trigonometry

We shall use the function

$$Q_n = \frac{\sin n\alpha}{\sin \alpha}, \tag{37}$$

where n is an integer and α is a non-zero real angle. In this paper trigonometric functions are to be considered as power series. It will be pointed out that in (37) we violate the taboo against division by a power series with a zero constant term. So let us write $\sin x = xS(x)$ and interpret the quotient in (37) as $nS(n\alpha)/S(\alpha)$.

Using the formula for the sine of a sum we can establish the recursion formula

$$Q_{r+2} = 2 \cos \alpha \cdot Q_{r+1} - Q_r. \tag{38}$$

From (37) we have $Q_0 = 0$ and $Q_1 = 1$. For other values of n we can calculate Q_n from (38). Thus, writing ξ for $2 \cos \alpha$ we have

$$Q_2 = \xi,$$

$$Q_3 = \xi^2 - 1,$$

$$Q_4 = \xi^3 - 2\xi,$$

$$Q_5 = \xi^4 - 3\xi^2 + 1,$$

$$Q_6 = \xi^5 - 4\xi^3 + 3\xi,$$

$$Q_7 = \xi^6 - 5\xi^4 + 6\xi^2 - 1.$$

More generally we can infer from (38) that if $n > 0$ then Q_n is a polynomial in ξ of degree $n - 1$ and the coefficient of its leading term is unity. Also the indices of ξ in the terms with non-zero coefficients all have the parity of $n - 1$.

The general formula is

$$Q_n = \sum_{j=0}^{m-1} \frac{(-1)^j (n - j - 1)! \xi^{n-2j-1}}{j!(n - 2j - 1)!}. \tag{39}$$

where m is the integral part of $(n + 1)/2$. One way of proving it is by induction, using (38).

In what follows we take n to be an even number. So $n = 2m$. Writing $\pi/2 - \alpha$ for α in (39) we transform to

$$\sin n\alpha = \cos \alpha \sum_{j=0}^{m-1} \frac{(-1)^{m+j-1}(n-j-1)!(2\sin\alpha)^{n-2j-1}}{j!(n-2j-1)},$$

whence by integration we have

$$\cos n\alpha = m \sum_{j=0}^{m} \frac{(-1)^{m+j}(n-j-1)!(2\sin\alpha)^{n-2j}}{j!(n-2j)!}$$

$$= m \sum_{j=0}^{M} \frac{(-1)^{j}(m+j-1)!(2\sin\alpha)^{2j}}{(2j)!(m-j)!}. \tag{40}$$

The polynomials Q_n ($n > 0$) are, with trivial adjustments, the "Beraha polynomials" of other papers on chromatic polynomials. See e.g. [11].

The zeros of the polynomial Q_n, given immediately by (37), correspond to the product formula

$$Q_n = \prod_{r=1}^{n-1} \left\{ 2\cos\alpha - 2\cos\frac{\pi r}{n} \right\}. \tag{41}$$

In earlier papers we often distinguish between the "even case" in which n is an even number $2M$ and the "odd case" in which n has the form $2M - 1$. Here, as mentioned above, we consider the even case only, expecting that this restriction will give us a shorter path from (α) to (ω). In the even case the zero at $r = M$ stands alone, but the other values of r come in pairs $\{r, n-r\}$ so that the zeros given by each pair are equal in magnitude but opposite in sign. Accordingly

$$\sin n\alpha = \sin \alpha \cdot \{2\cos\alpha\} \prod_{r=1}^{M-1} \left\{ 4\cos^2\alpha - 4\cos^2\frac{\pi r}{n} \right\}, \tag{42}$$

or equivalently

$$\sin n\alpha = 2^{M-1} \cdot \sin 2\alpha \cdot \prod_{r=1}^{M-1} \left\{ \cos 2\alpha - \cos\frac{2\pi r}{n} \right\}. \tag{43}$$

We need also the result of substituting $\pi/2 - \alpha$ for α in (43). This substitution interchanges $\sin\alpha$ and $\cos\alpha$ and replaces $\cos 2\alpha$ by $-\cos 2\alpha$. It replaces $\sin n\alpha$ by

$(-1)^{M-1}\sin n\alpha$ in the even case. So we have the following equation:

$$\sin 2M\alpha = 2^{M-1} \cdot \sin 2\alpha \prod_{r=1}^{M-1} \left\{ \cos 2\alpha + \cos \frac{\pi r}{M} \right\}, \tag{44}$$

10. Some special invariants

It is now time to make use of the variable V, given in terms of y by equation (13). When y is replaced in (13) by y_1 or y_2 then V is changed into a member, V_1 or V_2 respectively, of $Y2(s, \lambda)$. Likewise we denote the result of substituting y_1 or y_2 for y in the invariant K by K_1 or K_2 respectively. Each of these is obtained from the corresponding J_i by adding v. They are of course equal members of $Y2(s, \lambda)$.

After the transformation from y to V the series $L(y, t, \lambda)$ becomes a function of V, t and λ. It is a member of $Y2(V, \lambda)$.

By a partial transformation from y to V in the definition of K, we find that

$$\lambda^{-1} t L(y, t, \lambda) - yt(1-y)^{-1} = K - V^2 + \mu v V - v^2. \tag{45}$$

Completing the transformation we get

$$K = (V^2 - \mu v V + v^2) + t \left\{ \frac{1}{v - V} + A \right\}, \tag{46}$$

where A is a member of $Y2(V, \lambda)$ in which the denominators of the coefficients P/Q are powers of $V - v + 1$. (See (13).)

Let i and j be 1 and 2 in some order. Let us now substitute y_i for y, and equivalently V_i for V, in (45). Eliminating the expression on the left by (25) and (26), and completing the conversion to V, we obtain

$$K_i - (V_i^2 - \mu v V_i + v^2) = (V_j - v)(V_j - \mu V_i + v). \tag{47}$$

Let us now assign the value $2\pi/N$ to the angle α of the definition of Q_n. Here N is a fixed positive even integer $2M$. We also assign the value $2 \cos \alpha = 2 \cos(2\pi/N)$ to μ at this stage. We are thus restricting ourselves to some special values of $\lambda = 2 + \mu$, all positive but most of them non-integral.

$$\alpha = \frac{2\pi}{N}, \qquad \xi = \mu = 2 \cos \frac{2\pi}{N}. \tag{48}$$

This substitution for μ converts K into an invariant $K(N)$ of $Y2(y)$.

We now take note of the identity

$$Q_m^2(V_j - v)(V_j - \mu V_i + v)$$

$$+ \{Q_{m-1}V_i + v(1 - Q_m)\}\{Q_{m+1}V_i - v(1 + Q_m)\}$$

$$+ \{Q_m V_j - Q_{m-1}V_i - v\}\{Q_{m+1}V_i - Q_m V_j - v)\} = 0, \qquad (49)$$

where m is any integer. To prove it we observe that the expression on the left is

$$V_i V_j \{-\mu Q_m^2 + Q_m Q_{m-1} + Q_m Q_{m-1}\} + V_1 v \{\mu Q_m^2 - Q_m Q_{m-1} - Q_m Q_{m+1})\}$$

which is zero, by (38) and (48).

For any positive integer r we write

$$E(r) = Q_r^2 \{K(N) - (V^2 - \mu v V + v^2)\}$$

$$+ \{Q_{r-1}V + v(1 - Q_r)\}\{Q_{r+1}V - v(1 + Q_r)\}. \qquad (50)$$

Evidently $E(r)$ is a member of $Y2(V)$. Substituting V_i for V in it we obtain $E_i(r)$. By (47) and (49) we have

$$E_i(r) = -\{Q_r V_j - Q_{r-1}V_i - v\}\{Q_{r+1}V_i - Q_r V_j - v\}. \qquad (51)$$

Hence, for any positive integer u,

$$(-1)^u \prod_{r=1}^{u} E_i(r) = \{V_j - v\}\{Q_{u+1}V_i - Q_u V_j - v\}C_u, \qquad (52)$$

where C is symmetrical in the suffixes i and j, that is 1 and 2.

We observe that now $Q_M = 0$ and $Q_{M-1} = 1$. We can therefore deduce the following theorem from (52).

THEOREM 10.1. *The product*

$$I(2M) = (v^2 - V^2) \prod_{r=1}^{M-1} E(r) \qquad (53)$$

is a special invariant.

From this theorem and the definition of $E(r)$, and with the help of (13) we see that $I(N)$ belongs to the set W_{M-2} defined in the preamble to Theorem 8.2 (if $M > 1$). So from Theorem 8.2 and equation (36) we have

THEOREM 10.2. *If $N > 2$ then*

$$I(N) = \sum_{i=0}^{M-2} g_i(K(N))^i, \tag{54}$$

where the g_i are power series in t with numerical coefficients.

This is the "exploitable equation" that we hoped to derive from the theory of invariants. The last part of [7] consists of a proof that our equations (53) and (54) can, theoretically, provide a recursive determination of $K(N)$ and each of the power series g_i. We need a reformulation of equation (53). That reformulation is the concern of the next section.

11. Trigonometrical formulae relating $I(N)$ to $K(N)$

By a rearrangement on the right of equation (50) we have

$$E(r) = Q_r^2 K(N) - V^2\{Q_r^2 - Q_{r-1}Q_{r+1}\} + vV\{Q_{r+1} - Q_{r-1}\} - v^2. \tag{55}$$

From (37), using elementary trigonometric identities, we can establish

$$Q_r^2 - Q_{r-1}Q_{r+1} = 1 \tag{56}$$

and

$$Q_{r+1} - Q_{r-1} = 2\cos r\alpha, \tag{57}$$

thus obtaining the following simplification of (55)

$$E(r) = Q_r^2 K(N) - V^2 + 2vV\cos r\alpha - v^2. \tag{58}$$

Now Q_r^2 can be written as $\{1 - \cos^2 r\alpha\}/\sin^2\alpha$. Substituting this in (58) and rearranging, we find that

$$-\frac{\sin^2\alpha \cdot E(r)}{K(N)} = \left\{\cos r\alpha - \frac{vV\sin^2\alpha}{K(N)}\right\} - \left\{1 - \frac{v^2\sin^2\alpha}{K(N)}\right\}\left\{1 - \frac{V^2\sin^2\alpha}{K(N)}\right\}. \tag{59}$$

In [8] we now encounter two functions θ and ϕ defined as inverse sines, thus:

$$\theta = \sin^{-1} \frac{V \cdot \sin \alpha}{\sqrt{K(N)}}, \qquad \phi = \sin^{-1} \frac{v \cdot \sin \alpha}{\sqrt{K(N)}}.$$

Equation (59) is then rewritten in terms of θ and ϕ, and the argument proceeds through a sequence of trigonometrical identities. But we can make two criticisms of this procedure.

First, when we introduce a square root of K, we must expect to go outside $Y2(V, \lambda)$. To meet this objection we define $Y3(V, \lambda)$ as the set of all members of $Y2(V, \lambda)$ together with the products of those members by the algebraic form

$$R = \sqrt{V^2 - \mu v V + v^2}, \tag{60}$$

R^2 being the coefficient of t^0 in $K(N)$. We say therefore that $K(N)$ has a square root in $Y3(V, \lambda)$, and we propose to work henceforward within that set of power series, restricted to our special λ. We note that $Y3(V, \lambda)$ is closed under addition and multiplication.

Second, it may be queried whether θ and ϕ, as defined above, are members of $Y2(V, \lambda)$ ore even $Y3(V, \lambda)$. For if we try to evaluate them as power series in t in the ordinary way we shall find the coefficients to be given by infinite sums. However there are ways round this difficulty. In what follows we introduce θ and ϕ as new variables, establish a polynomial identity in the quantities $\sin \theta$ and $\sin \phi$, and note that this identity must remain valid when we substitute $V \sin \alpha / \sqrt{K(N)}$ and $v \sin a / \sqrt{K(N)}$, or any other two members of $Y3(V, \lambda)$, for $\sin \theta$ and $\sin \phi$ respectively. In analogy with (59) we write

$$A(\theta, \phi, r) = \{\cos r\alpha - \sin \theta \sin \phi\}^2 - (1 - \sin^2 \theta)(1 - \sin^2 \phi),$$

$$= \{\cos r\alpha - \cos(\theta - \phi)\}\{\cos r\alpha + \cos(\theta + \phi)\}. \tag{61}$$

noting that, by (59), the proposed substitution transforms $A(\theta, \phi, r)$ into

$$-\frac{\sin^2 \alpha E(r)}{K(N)}.$$

We now recall that $\alpha = 2\pi/N$, N being as before an even integer exceeding 3. With this restriction we apply (43) and (44). But for the "α" of those equations we substitute $\theta - \phi$ or $\theta + \phi$ as appropriate.

$$\prod_{r=1}^{M-1} A(\theta, \phi, r) = \frac{(-1)^{M-1} \sin M(\theta - \phi)}{2^{M-1} \sin(\theta - \phi)} \cdot \frac{\sin M(\theta + \phi)}{2^{M-1} \sin(\theta + \phi)}$$

$$= \frac{(-1)^{M-1} \{\cos 2M\theta - \cos 2M\phi\}}{4^{M-1} \{\cos 2\theta - \cos 2\phi\}}. \tag{62}$$

Now $\cos 2M\theta$ is a polynomial in $\sin \theta$. The form of this polynomial has been noted in (40). Hence (62), after multiplication by the final denominator on the right, is a polynomial identity in $\sin \theta$ and $\sin \phi$, and we may legitimately make the proposed substitution in it. Using (53) we get the following equation:

$$\frac{1}{2} \left\{ \frac{4 \sin^2(2\pi/N)}{(K(N))} \right\}^M I(2M) = \cos\left\{ 2M \sin^{-1} \frac{V \sin(2\pi/N)}{\sqrt{K(N)}} \right\}$$

$$- \cos\left\{ 2M \sin^{-1} \frac{v \sin(2\pi/N)}{\sqrt{K(N)}} \right\}. \tag{63}$$

In order to stay within $Y3(V, \lambda)$ we should, when θ is a member of that set, treat the function $\cos\{2M \sin^{-1} \theta\}$ as a polynomial in θ, not presuming to evaluate the formal inverse sine as the first of two stages.

12. Elimination of y_1 and y_2

The theory of invariants gives us the "exploitable" equation of Section 10. With this equation the functions y_1 and y_2, with the variable s, disappear. We perhaps gain by now having only the one function L of three variables, even though we have introduced $M - 2$ new functions g_i of two variables each. We seek now a convenient form for the "exploitable" equation.

Eliminating $I(N)$ from (54) and (63) we find

$$\cos\left\{ N \sin^{-1} \frac{V \sin(2\pi/N)}{\sqrt{K(N)}} \right\} - \cos\left\{ N \sin^{-1} \frac{v \sin(2\pi/N)}{\sqrt{K(N)}} \right\}$$

$$= \frac{1}{2} \left(4 \sin^2(2\pi/N)^M \sum_{j=0}^{M-1} g_j (K(N))^{j-M} \right). \tag{64}$$

We now apply (40), multiply by $(K(N))^M$, and divide by $(2 \sin(2\pi/N))^N$.

$$M \sum_{j=0}^{M} \frac{(-1)^{M+j}(N-j-1)!(V^{N-2j} - v^{N-2j})(K(N))^j}{j!(N-2j)!(4\sin^2(2\pi/N))^j} = \frac{1}{2} \sum_{j=0}^{M-2} g_j(K(N))^j. \quad (65)$$

Here (65) is the "exploitable" equation.

We do not attempt here a complete solution of (65). But we do use it to get formulae for g_{M-2} and g_{M-3}, formulae that are needed in the following sections. By (45) we have (for general λ)

$$K = (V^2 - \mu v V + v^2) + \lambda^{-1} t L(y, t, \lambda) + \frac{t}{v - V}. \quad (66)$$

Substitution of v for V transforms $V^2 - \mu v V + v^2$ into v. Applied to $\lambda^{-1} t L(y, t, \lambda)$ it replaces y by 1, giving $\lambda^{-1} t h$. Here h is the function of t and λ, defined in Sections 1 and 2. So, for each positive integer r, we can write

$$K^T = \left\{ \frac{t}{v - V} \right\}^r + r(\lambda^{-1} t h + v) \left\{ \frac{t}{v - V} \right\}^{r-1} + (v - V)B_{r-1}. \quad (67)$$

Here B_j is our symbol for a member of $Y2(V, \lambda)$, not necessarily the same member at each occurrence, in which no denominator of a power of t divided by a power of $v - V$ higher than the jth. In a similar way we can write

$$V^{2r} = v^{2r} - 2rv^{2r-1}(v - V) + r(2r - 1)v^{2r-2}(v - V)^2 + B_0(v - V)^3. \quad (68)$$

Substituting from (67) and (68) in (65), we get the following abridgement of (65), one that retains all the information necessary for our immediate purpose:

$$\frac{M^2}{2(4 - \mu^2)t^{M-1}} \cdot \left\{ \frac{2vt^{M-1}}{(v-V)^{M-1}} + \frac{2v(M-1)(\lambda^{-1}ht + v)t^{M-2}}{(v-V)^{M-3}} - \frac{t^{M-1}}{(v-V)^{M-3}} \right\}$$

$$- \frac{M^2(M^2 - 1) \cdot 4v^3 t^{M-2}}{4!(4 - \mu^2)^{M-2}(v-V)^{M-3}}$$

$$= \frac{t^{M-2}g_{M-2}}{2(v-V)^{M-2}} + \frac{(M-2)(\lambda^{-1}th + v)g_{M-2}t^{M-3}}{2(v-V)^{M-3}}$$

$$+ \frac{t^{M-3}g_{M-3}}{2(v-V)^{M-3}} + (n - V)B_{M-3}. \quad (69)$$

Equating terms involving the $(M-2)$th power of $v - V$ we find that

$$(4 - \mu^2)^{M-1} g_{M-2} = 2vM^2 t. \tag{70}$$

Next we equate terms in $(v - V)^{M-3}$. Using (70) we find that

$$g_{M-3} = \frac{2vM^2 t(\lambda^{-1} th + v)}{(4 - \mu^2)^{M-1}} - \frac{M^2 t^2}{(4 - \mu^2)^{M-1}} - \frac{M^2(M^2 - 1)v^3 t}{3(4 - \mu^2)^{M-2}}. \tag{71}$$

We might continue this procedure, seeking formulae for g_{M-4}, g_{M-5} and so on. But we content ourselves with the following general observation.

THEOREM 12.1. *Each of the power series g_j in t has a zero constant term.*

This theorem follows from the fact that each term on the right of (65) divides by $v - V$. Each term $\{t/(v - V)\}^r$, with $r > 0$, in the full expansion of K^j becomes $t\{t/(v - V)\}^{r-1}$ in $(v - V)K^j$. It follows that when the left side is expanded in the form $\sum m_j \{t/(v - V)\}^j$, with m_j independent of V, then each m_j must divide by t. Equating the coefficients of like powers of $t/(v - V)$ on the two sides of (67) then establishes the theorem.

13. An auxiliary functional equation

In equation (63) the term involving the cosine of

$$N \sin^{-1} \frac{V \sin \alpha}{\sqrt{K(N)}}$$

could be reduced to triviality if we were permitted to put

$$\frac{V \sin \alpha}{\sqrt{K(N)}} = \sin \frac{\pi r}{2N} \tag{72}$$

for some integer r.

In this section we find that this restriction can be enforced, for some positive integral values of r, by substituting for V a suitable power series f of the form

$$f = f_0 + f_1 t + f_2 t^2 + f_3 t^3 + \cdots, \tag{73}$$

where the f_j are real numbers. We denote the class of all such power series by $Y2(-)$. By the substitution the function $K(N)$ of V is replaced by a member $K(N, f)$ of $Y2(-)$. We recall that $K(N)$ has coefficients whose denominators are powers of $V - v$ and $V - v + 1$. So by the convention adopted in Section 4 we must reject, as a proposed f, any member of $Y2(-)$ whose constant term is v or $v - 1$.

We shall use the following existence theorem.

13.1. Let b be a real number that is not 0, v, $v - 1$ or $2v/\mu$. Then there exists a unique power series f in $Y2(-)$ such that $f_0 = b$ and

$$f^2 = cK(N, f),\tag{74}$$

where

$$c = \frac{b^2}{b^2 - \mu vb + v^2}.\tag{75}$$

We note that (75) is a consequence of (74), obtained from it by equating constant terms. Before proving 13.1 we make the following further observation.

13.2. The number c of Theorem 13.1 is positive.

To prove 13.2 we observe that the denominator in (75) can be written as

$$\{b - \mu v/2\}^2 + v^2\{1 - \mu^2/4\}.$$

But $1 - \mu^2/4$ is positive, since μ now stands for $2 \cos 2\pi/N$.

In our proof of 13.1 we use a symbol A_s defined as standing for any known polynomial in those coefficients f_j for which j is less than the positive integer s. "Known" means calculable in terms of the coefficients of $K(N)$. The symbol A_s need not denote the same polynomial at each appearance.

Suppose for example that we ask for the coefficient k_s of t^s in $K(N, f)$. To find the contribution to k_s from the coefficient K_j of t^j in K, where j is at most s, we must substitute f for V in K_j and then take the coefficient U_{s-j} of t^{s-j} in the result. We can accordingly write the contribution as an A_s unless $j = 0$. So we have

$$k_s = \{\text{Coeff. } t^s \text{ in } f^2 - \mu vf + v^2) + A_s, \qquad \text{by (46)}$$

$$= 2f_0 f_s - \mu vf_s + A_s.$$

Suppose as an inductive hypothesis that we have determined f_j, for all $j < s$, so that (74) holds up to the $(s - 1)$th power of t. By equating coefficients of t^2 on the two sides of (74) we find that the equation will hold up to the sth power of t if and only if we choose f_s so as to satisfy the equation

$$2f_0 f_s = c\{2f_0 f_s - \mu v f_s\} - A_s,$$

that is,

$$f_s\{2f_0 - 2f_0 c - \mu vc\} = A_s, \tag{76}$$

where A_s is "known". But (76) determines f_s uniquely unless

$$2f_0 - 2f_0 c + \mu vc = 0. \tag{77}$$

Suppose (77) to hold. Then, since $f_0 = b$, we have

$$2b\{b^2 - \mu vb + v^2\} = \{2b - \mu v\}b^2,$$

that is,

$$-2\mu vb + 2v^2 = -\mu vb, \qquad \text{by (75).}$$

But then $b = 2v/\mu$, contrary to the hypothesis of 13.1.

We conclude that, once f_0 is fixed as b, there is a procedure for calculating the other coefficients in f, in succession, and that this procedure determines them uniquely. The proof of 13.1 is complete.

Suppose now that we decide upon a value for c and we ask for the corresponding value or values of b, if any. Then we need to solve the quadratic

$$b^2(c - 1) - \mu vbc - v^2 c = 0. \tag{78}$$

For later convenience we choose c to satisfy

$$\sqrt{c} = \frac{\sin \theta}{\sin(2\pi/N)}. \tag{79}$$

where θ is a real angle. We now divide (78) by b^2 and solve the quadratic for $1/b$. We find that

$$\frac{2v}{b} = \mu + \sqrt{4 - \mu^2} \cot \theta, \tag{80}$$

that is,

$$\frac{v \sin \theta}{b} = \sin\left(\theta \pm \frac{\pi}{M}\right). \tag{81}$$

We restrict θ to the range $0 < \theta \leq \pi/2$. We note that in this range $\cot \theta$ is non-negative and strictly decreasing. We can therefore be sure that in this range different values of θ correspond to different values of b. Moreover, there is no value of θ in the chosen range that makes b infinite; for this to happen $\cot \theta$ would have to be negative, by (80).

For later convenience we note that, with positive radicals,

$$\mu = 2 \cos \frac{2\pi}{N}, \qquad \sqrt{4 - \mu^2} = 2 \sin \frac{2\pi}{N},$$

$$\sqrt{2 - \mu} = 2 \sin \frac{\pi}{N}, \qquad \sqrt{2 + \mu} = 2 \cos \frac{\pi}{N},$$

$$v = \frac{1}{2 - \mu} = \frac{1}{4 \sin^2 \pi/N}.$$

We have excluded the values v, $v - 1$, and $2v/\mu$ of b. Let us investigate the values of θ to which these would correspond.

Suppose first that $b = v$. Then by (81) we have

$$\sin \theta = \sin\left(\theta + \frac{\pi}{M}\right), \qquad \theta = \frac{(M - 1)\pi}{N}.$$

For $b = 2v/\mu$ we find that $\cot \theta = 0$, by (80), whence $\theta = \pi/2$.

The value $v - 1$ of b is found to correspond to

$$\theta = \frac{(M - 3)\pi}{N}.$$

To show this we substitute that value of θ in (81). Then

$$b \cos \frac{\pi}{N} = v \cos \frac{3\pi}{N},$$

$$b = v \left\{ \cos \frac{2\pi}{N} - 2 \sin^2 \frac{\pi}{N} \right\}$$

$$= \frac{\mu v - 1}{2} = \frac{v(2 - v^{-1}) - 1}{2} = v - 1.$$

For a reason given in the next section we find it necessary to exclude also the value $b = \mu v/2$. We need to know that this corresponds to

$$\theta = \frac{(M - 2)\pi}{N}.$$

To prove this we substitute that value of θ in (81), finding that

$$b = v \cos \frac{2\pi}{N} = \frac{\mu v}{2}.$$

From now on $\mu v/2$ is our fourth unacceptable non-zero value of b.

We now propose to restrict θ to the form $\pi r/N$, where r is an integer. The expression on the right of (81) can now be written as

$$\sin \frac{(r + 2)\pi}{N} \qquad \text{or} \qquad \sin \frac{(r - 2)\pi}{N}$$

according as the positive or negative sign is used. Let us write S for the chosen integer $r + 2$ or $r - 2$. We now ask what values of S, in the range $1 \leq S \leq M$ are "admissible", that is corresponding to acceptable values of b.

THEOREM 13.1. *If $M \geq 8$ then all values of S from 1 to $M - 2$ are admissible, but the values $M - 1$ and M are not.*

Proof. Putting $r = 3$, 4 in (81) we can arrange that $S = 1$, 2 by using the negative sign. Giving r the values $1, 2, 3, \ldots, M - 4$ and using the positive sign we get the values $3, 4, 5, \ldots, M - 2$ of S. Since $M \geq 8$ we have used no r that corresponds to an excluded value of b.

To get $S = M - 1$ we would have to put r equal to $M - 3$ or $M + 1$, the latter value being equivalent to $M - 1$. But $M - 1$ and $M - 3$ are unacceptable values of r.

To get $S = M$ we would have to put r equal to $M - 2$ or $M + 2$, the two values being equivalent. But $M - 2$, corresponding to the value $\mu v/2$ or b, is an unacceptable value of r.

The above proof fails for $M < 8$ since at least one of the values 1, 3 and 4 of r is then not acceptable. In [9] one purports to prove, in effect, that there are $M - 2$ admissible values of S for all values of M from 3 upwards. But this proof now seems doubtful to the author. Fortunately Theorem 13.1 contains enough information to get us to equation (ω), even though that equation is asserted for general λ.

For a given acceptable r let us consider the substitution of the corresponding f for V in some expressions involving V.

$$K(N) \text{ becomes } f^2/c, \qquad \text{by (74)}, \tag{82}$$

$$\frac{V \sin \alpha}{\sqrt{K(N)}} \text{ becomes } \sin \frac{\pi r}{N}. \tag{83}$$

$$\frac{2v \sin \alpha}{\sqrt{K(N)}} \text{ becomes } \frac{2v \sin(\pi r/N)}{f}. \tag{84}$$

We now write

$$w_S = \frac{4 \sin^2(\pi r/N)}{f^2}, \tag{85}$$

using the f and r that correspond to the chosen S. We call w_S our "parameter of order S". The expression on the right of (84), divided by v, is a square root of w_S. By (81) the values of S from 1 to $M - 2$ correspond to $M - 2$ distinct parameters w_S. For on the left we have the constant term in the halved square root of w_S, and on the right we have $\sin(\pi S/N)$.

14. Parametric equations

Using (40) we rewrite (64) thus:

$$\cos\left\{N \sin^{-1} \frac{V \sin 2\pi/N}{\sqrt{K(N)}}\right\} = \sum_{j=0}^{M} p_j \left\{\frac{4 \sin^2(2\pi/N)}{K(N)}\right\}^j, \tag{86}$$

where

$$p_j = M \frac{(-1)^j (M+j-1)! v^{2j}}{(M-j)!(2j)!} + g_{M-j}\{4 \sin^2(2\pi/N)\}^{M-j}. \qquad (87)$$

In (87) we put $g_M = g_{M-1} = 0$.

Now let us substitute for V in (86) the series f corresponding to an admissible integer r and the corresponding S. We get

$$(-1)^r = (-1)^S = \sum_{j=0}^{M} p_j w_S^j. \qquad (88)$$

Our next step is to differentiate (86) by V and then substitute f for V. On the left the rule for differentiating a function of a function gives us a multiple of

$$\sin\left\{ N \sin^{-1} \frac{V \sin(2\pi/N)}{\sqrt{K(N)}} \right\}.$$

So after the substitution the left side of the equation takes the value zero. This time it is left to the reader to decide whether the various steps of the differentiation should be kept within $Y3(V)$, and how this can be done using the fact that $\cos N\theta$ and $\sin N\theta/\cos\theta$ are polynomials in $\sin\theta$.

In what follows we use the symbol D_x to denote partial differentiation with respect to a variable x.

After differentiation and substitution the right side of (86) becomes a multiple of

$$\sum_{j=0}^{M} j p_j w_S^{j-1},$$

and a non-zero multiple, provided we can be assured that $D_V(K(N))$ does not become identically zero when f is substituted for V. The constant term in $D_V(K(N))$ after the substitution can be calculated from (46) as $2b - \mu v$. So we have no problem for any acceptable value of r. However the constant term becomes zero when $r = M - 2$ and $S = M$, that is when $b = \mu v/2$. In that case the possibility that $D_V(K(N))$ becomes zero after the substitution is not excluded. That is our reason for rejecting the value $\mu v/2$ of b. The actual possibilities for $r = M - 2$ and $S = M$ can be stated as follows.

THEOREM 14.1. *Let $r = M - 2$ and $S = M$, and consider the corresponding f. Then either*

$$[D_V(K(N))]_{V=f} = 0$$

or there exists a corresponding parameter w_M distinct from each of the $M - 2$ already recognized.

Since we are avoiding the value $M - 2$ of r we can now complement (88) with the following equation, valid for all S from 1 to $M - 2$,

$$0 = \sum_{j=1}^{M} j p_j w_S'^{j-1}. \tag{89}$$

We may expect the $2M - 4$ parametric equations (88) and (89) to determine the $M - 2$ parameters w_S, together with the $M - 2$ unknown quantities p_j. (p_0, p_1 and p_2 are known in terms of h, by (87), (70) and (71).) The paper [9] purports to give examples for some small values of M, less than 8. The procedure is valid, provided that the necessary $M - 2$ parameters can be shown to exist. For $M = 3$ the one parameter needed is given by $r = -1$ and $S = 1$.

15. Partial differential equations

We introduce a new indeterminate u. Denoting its square by w and working in $Y2(u)$ we write

$$U = \sum_{j=0}^{M} p_j w^j. \tag{90}$$

Then

$$D_w(U) = \sum_{j=1}^{M} j p_j w^{j-1} \tag{91}$$

and

$$D_t(U) = \sum_{j=2}^{M} \frac{dp_j}{dt} w^j, \qquad \text{by (89).} \tag{92}$$

Our $M - 2$ distinct parameters are roots of the polynomial $U^2 - 1$ in w, by (88). And they are also roots of the polynomial $D_w(U)$, by (89) and (91). It follows that they are double roots of $U^2 - 1$. Hence they are roots of

$$D_t(U^2 - 1) = 2UD_t(U).$$

Since no parameter is a root of U, by (88), it follows that all our $M - 2$ parameters are roots of $D_t(U)$. We infer from these observations that

$$wD_w(U) = (\alpha + \beta w)D_t(U), \tag{93}$$

where α and β are independent of u, functions of t alone. We infer also that

$$w^4(U^2 - 1) = Y\{D_t(U)\}^2, \tag{94}$$

where Y is a polynomial of degree 4 in w. Now $p_0 = 1$ and p_2 is non-zero, by (87). Hence, by (90) and (92), Y must divide by w. We can therefore write

$$Y = w\{A + Bw + Cw^2 + Dw^3\}, \tag{95}$$

where A, B, C and D are independent of u. By (90) and (94) the coefficient A is positive.

We do not have to accept α, β and the coefficients in Y as entirely new and unknown functions of t. Actually they can all be expressed in terms of h. We proceed to do this for α and β, leaving the study of A, B, C and D to the next section.

Equating coefficients, first of w^2 and then of w^3, in (93) we find that

$$p_1 = \alpha \frac{dp_2}{dt}, \tag{96}$$

$$2p_2 = \beta \frac{dp_2}{dt} + \alpha \frac{dp_3}{dt}. \tag{97}$$

From (87), (70) and (71) we find

$$p_1 = -\frac{M^2 v^2}{2},\tag{98}$$

$$p_2 = \frac{M^2 vt}{4 - \mu^2} + \frac{M^2 (M^2 - 1) v^4}{4!},\tag{99}$$

$$\frac{dp_2}{dt} = \frac{M^2 v}{4 - \mu^2},\tag{100}$$

$$\frac{dp_3}{dt} = \frac{M^2 v}{(4 - \mu^2)^2} \cdot \frac{d(\lambda^{-1} t^2 h + vt)}{dt} - \frac{M^2 t}{(4 - \mu^2)^2} - \frac{M^2 (M^2 - 1) v^3}{3!(4 - \mu^2)}.\tag{101}$$

Applying (98) and (100) to (96) we find that

$$\alpha = -v(4 - \mu^2)/2 = -(2 + \mu)/2 = -\lambda/2.\tag{102}$$

Applying (99) and (101) to (97) we get

$$2\beta = 3t + v \frac{d\{\lambda^{-1} t^2 h + vt\}}{dt}.\tag{103}$$

In [10] the left side of the corresponding equation is written as "β" instead of "2β". This is because in the equation corresponding to our (91) the differentiation is with respect to our u, not w.

We can take square roots in (94) to get

$$w^2 \sqrt{\frac{U^2 - 1}{Y}} = D_t(U).\tag{104}$$

(Note that this operation can be carried out within $Y2(u)$.) Then we can apply (93) to get

$$(\alpha + \beta w) \sqrt{\frac{U^2 - 1}{Y}} = D_w(U).\tag{105}$$

We now differentiate (104) by w and (105) by t. We can then equate the two left sides. In the resulting equation each side has a term involving a derivative of U. But the two terms are equal by (93), and can be deleted. From the remaining terms we can cancel the common factor $\sqrt{U^2 - 1}$ to get a partial differential equation for the single unknown function Y, as follows

$$(\alpha + \beta w)D_t(Y) - w^2 D_w(Y) + 2w Y(2 - \beta') = 0, \tag{106}$$

a prime being used, here an in what follows, to denote differentiation with respect to t.

The polynomial in w on the right of equation (92) is of degree M and has a double zero at $w = 0$. We have recognized the $M - 2$ other zeros in our $M - 2$ distinct parameters. Since the polynomial can have no other zero, we infer that there can be no $(M - 1)$th parameter, and so rule out one of the alternatives in Theorem 14.1. We thus establish the substitutional differential equation there presented. (The author, thinking this an odd way to prove a differential equation, has made a check up to the third power of t.) The equation of 14.1 appears in this paper only as a curiosity met with on the way from (α) to (ω).

16. A differential equation

From (95) and (106) we have

$$(\alpha + \beta w)(A'w + B'w^2 + C'w^3 + D'w^4) - (Aw^2 + 2Bw^3 + 3Cw^4 + 4Dw^5)$$
$$+ (4 - 2\beta')(Aw^2 + Bw^3 + Cw^4 + Dw^5) = 0. \tag{107}$$

By equating coefficients of like powers of w in (107) we obtain the following equations.

$$\alpha A' = 0, \tag{108}$$

$$(3 - 2\beta')A + \alpha B' + \beta A' = 0, \tag{109}$$

$$(2 - 2\beta')B + \alpha C' + \beta B' = 0, \tag{110}$$

$$(1 - 2\beta')C + \alpha D' + \beta C' = 0, \tag{111}$$

$$-2\beta' D + \beta D' = 0. \tag{112}$$

In order to integrate these equations we need to know the constant terms A_0, B_0, C_0 and D_0 in the functions A, B, C and D respectively. Let us also write b_0, U_0

and $Y0$ for the coefficients of t^0 in b, U and Y respectively. We use equation (40) and (87) and Theorem 12.1 to show that

$$U_0 = \cos N \sin^{-1}(vu/2). \tag{113}$$

Taking coefficients of t^0 in (103) we can calculate that

$$2\beta_0 = v^2. \tag{114}$$

We next square both sides of (105) and take coefficients of t^0. We then substitute from (113), cancel the common factor and find that, after rearrangement,

$$(\alpha + \beta_0 w)^2(4 - v^2 w) = N^2 v^2(A_0 + B_0 w + C_0 w^2 + D_0 w^3). \tag{115}$$

By equating coefficients of the various powers of w we get the following equations:

$$4N^2 v^2 A_0 = 16\alpha^2 = 4\lambda^2, \tag{116}$$

$$4N^2 v^2 B_0 = 32\alpha\beta_0 - 4v^2\alpha^2 = -8\lambda v^2 - \lambda^2 v^2, \tag{117}$$

$$4N^2 v^2 C_0 = 16\beta_0^2 - 8\alpha\beta_0 v^2 = 4v^4 + 2\lambda v^4, \tag{118}$$

$$4N^2 v^2 D_0 = -4v^2\beta_0^2 = -6. \tag{119}$$

We proceed to eliminate A, B, C and D from equations (108) to (112). We make use of the function

$$\gamma = \frac{3t^2}{4} + \frac{v}{2}\{\lambda^{-1}t^2 h + vt\}, \tag{120}$$

which is an integral of β, by (103). Let us also write T for $4N^2 v^2$.
Integrating (108), using (116) we get

$$TA = 4\lambda^2. \tag{121}$$

We multiply (109) by T and substitute from (107), getting

$$T\lambda B' = 2(3 - 2\beta') \cdot 4\lambda^2,$$

whence

$$TB = 8\lambda(3t - 2\beta) + Z,$$

where Z is independent of t. By (103) and (117) we have

$$-8\lambda v^2 - \lambda^2 v^2 = -\delta\lambda v^2 + Z,$$

$$TB = 24\lambda t - 16\lambda\beta - \lambda^2 v^2. \tag{122}$$

A similar treatment of (112) yields the formula

$$TC = 4\lambda v^2 t - 48t^2 - 40\gamma + 96\beta t - 4\lambda v^2\beta - 16\beta^2. \tag{123}$$

Solving the differential equation (112) with the help of (119), we find that

$$TD = 16v^2\beta^2. \tag{124}$$

We have not yet used equation (111). Substituting the above formulae for A, B, C and D in it and simplifying we obtain the following differential equation for γ.

$$(1 - 2\gamma'')(-\lambda v^2 t + 12t^2 + 20\gamma) + 24\gamma'\gamma''t = 0. \tag{125}$$

We can regard (125) as a differential equation for h. Here we transform it into an equation for the function $H = t^2h$ introduced in Section 1. We substitute for γ and its derivatives in accordance with (103) and (120), finding that

$$\{2 + \lambda^{-1}vH''\}\{27t^2 - \lambda v^2 t + 10v^2 t + 10\lambda^{-1}vH\}$$

$$-6t\{3t + v^2 + \lambda^{-1}vH'\}\{3 + \lambda^{-1}vH''\} = 0. \tag{126}$$

We now suppose the left side of (126) to be multiplied out and we collect the terms formally involving like powers of v, as though λ and v were independent variables.

The coefficient of v^3 is

$$\lambda^{-1}H''\{-\lambda t + 10t\} - 6\lambda^{-1}H''t = \lambda^{-1}H''t(4 - \lambda).$$

But $(4 - \lambda)$ is the reciprocal of v. Hence the terms formally involving v^3 actually sum to $\lambda^{-1}H''tv^2$. At a later stage we will combine this sum with the terms formally involving v^2.

The coefficient of v^2 is

$$2\{-\lambda t + 10t\} + 10\lambda^{-2}HH'' - 6t\{3 + \lambda^{-2}H'H''\}$$
$$= -2\lambda t + 2t + 10\lambda^{-2}HH'' - 6\lambda^{-2}H'H''t.$$

The coefficient of v^1 is

$$2\{10\lambda^{-1}H\} + \lambda^{-1}H'' \cdot 27t^2 - 6t\{3t\lambda^{-1}H'' + 3\lambda^{-1}H'\}$$
$$= 20\lambda^{-1}H + 9\lambda^{-1}t^2H'' - 18\lambda^{-1}H't.$$

The coefficient of v^0 is

$$2 \cdot 27t^2 - 6 \cdot 3t \cdot 3 \cdot t = 0.$$

We can now rewrite (126), after multiplying twice by λ and twice by $4 - \lambda$, as

$$H''\{\lambda t + 10H - 6H't\} - 2\lambda^3 t + 2\lambda^2 t + \lambda(4 - \lambda)\{20H + 9t^2H'' - 18H't\} = 0.$$

$$(127)$$

We can assert that this equation for H is valid whenever μ, that is $\lambda - 2$, is of the form $2\cos(\pi/M)$, where M is an integer not less than 8. (See Theorem 13.1.)

17. General λ

Let us discuss equation (127), altering its interpretation somewhat by regarding λ as an independent indeterminate. We ask if it is possible, by the technique of solution in series to construct a solution

$$H = H(0) + H(1)t + H(2)t^2 + H(3)t^3 + \cdots$$

in which $H(0) = H(1) = 0$.

We begin the construction by taking coefficients of t^1 in (127). Only the terms $H''\lambda t$, $-2\lambda^3 t$ and $\lambda^2 t$ contribute and from them we see that $H(2)$ must be $\lambda(\lambda - 1)$. Now suppose we have determined coefficients in the attempted solution up to $H(k)$, where $k > 2$. We try to find $H(k + 1)$ by equating coefficients of t^k in (127). The first of the three terms just given special mention contributes the coefficient $2H(k + 1)$. The rest of the formula gives a polynomial in λ and the known coefficients $H(m)$ with $m \leq k$. Having made this observation we can easily show by induction that the required solution H exists, that its coefficients for powers of t are uniquely determined and that they are polynomials in λ with integral coefficients. In the notation of Section 4 the solution H belongs to $Y1(\lambda)$.

In Section 2 we have defined another H in $Y1(\lambda)$, whose coefficients are chromatic sums. The argument of the preceding Sections has shown that the two H's become equal for infinitely many values of λ, those noted at the end of the preceding section.

In particular the coefficients of any particular power t^k in the two H's become equal at all these special values of λ. Since the two coefficients are known to be polynomials in λ it follows that they are identical. We may therefore assert

THEOREM 17.1. *The differential equation* (127) *has a unique solution in* $Y1(\lambda)$, *subject to the condition that* $H(0) = H(1) = 0$. *This solution is the chromatic generating function* H *of Section* 2.

The equation of Theorem 14.1, finally established in the last paragraph of Section 15, can be extended to general λ in a similar way.

18. Special values of λ

By Theorem 17.1 the non-linear differential equation (127) determines H uniquely for any real or complex value of λ. We can put λ equal to 0, 1 or 2, finding that H is 0, 0 or $2t^2$ respectively, as it ought to be by the definitions of chromatic polynomials and near-triangulations. Besides these trivialities let us note that there are explicit formulae for the derivatives of g, q, L and h with respect to λ at $\lambda = 1$ and $\lambda = 2$ [3]. H has special algebraic properties when $\lambda = 2 + 2\cos(2\pi/N)$, where N is a positive integer. We have noted this in Section 14 for the even case $N = 2M$, and with the restriction that M is at least 8. The odd case $N = 2M - 1$ is discussed in some earlier papers ([7], [8], [9]). There are even explicit solutions for H in the cases $N = 5$ and $N = 6$ ([4], [5]); also for the case of infinite λ as this is interpreted in [6].

In the case $\lambda = 4$ equation (127) reduces to equation (ω), repeated here:

$$H''\{2t + 5H - 3H't\} = 48t. \qquad (\omega)$$

The author was glad to learn that this equation has been found to be of interest in itself, apart from its chromatic connection. Its number-theoretical and analytic properties are now under investigation by S. Beraha, J. Diamond and colleagues [1].

REFERENCES

[1] BERAHA, S., *Private communication.*
[2] TUTTE, W. T., *A census of planar triangulations,* Canad. J. Math. *16* (1962), 21–38.
[3] TUTTE, W. T., *Chromatic sums for rooted planar triangulations; the cases $\lambda = 1$ and $\lambda = 2$.* Canad. J. Math. *25* (1973), 426–447.
[4] TUTTE, W. T., *Chromatic sums for rooted planar triangulations II; the case $\lambda = \tau + 1$.* Canad. J. Math. *25* (1973), 657–671.
[5] TUTTE, W. T., *Chromatic sums for rooted planar triangulations III; the case $\lambda = 3$.* Canad. J. Math. *25* (1973), 780–790.
[6] TUTTE, W. T., *Chromatic sums for rooted planar triangulations IV; the case $\lambda = \infty$.* Canad. J. Math. *25* (1973), 929–940.
[7] TUTTE, W. T., *Chromatic sums for rooted planar triangulations V; special equations.* Canad. J. Math. *26* (1974), 893–907.
[8] TUTTE, W. T., *On a pair of functional equations of combinatorial interest.* Aequationes Math. *17* (1978), 121–140.
[9] TUTTE, W. T., *Chromatic solutions.* Canad. J. Math. *34* (1982), 741–758.
[10] TUTTE, W. T., *Chromatic solutions II.* Canad. J. Math. *34* (1982), 952–960.
[11] TUTTE, W. T., *The matrix of chromatic joins.* J. Combin. Theory Ser. B *57* (1993), 269–288.

Department of Combinatorics and Optimization,
University of Waterloo,
Waterloo, Ontario N2L 3G1,
Canada.

Aequationes Mathematicae **50** (1995) 135–142
University of Waterloo

0001–9054/95/020135–08 $1.50 + 0.20/0

Conditional functional equations and orthogonal additivity

LUIGI PAGANONI and JÜRG RÄTZ

Summary. Some examples of classes of conditional equations coming from information theory, geometry and from the social and behavioral sciences are presented. Then the classical case of the Cauchy equation on a restricted domain Ω is extensively discussed. Some results concerning the extension of local homomorphisms and the implication "Ω-additivity implies global additivity" are illustrated. Problems concerning the equations

$$[cf(x + y) - af(x) - bf(y) - d][f(x + y) - f(x) - f(y)] = 0$$

$$[g(x + y) - g(x) - g(y)][f(x + y) - f(x) - f(y)] = 0$$

$$f(x + y) - f(x) - f(y) \in V \qquad \text{(a suitable subset of the range)}$$

are presented.
The consideration of the conditional Cauchy equation is subsequently focused on the case when it makes sense to interpret Ω as a binary relation (orthogonality):

$$f: (X, +, \perp) \to (Y, +); \qquad f(x + z) = f(x) + f(z) \qquad (\forall x, z \in Z; x \perp z).$$

A brief sketch on solutions under regularity conditions is given. It is then shown that all regularity conditions can be removed. Finally, several applications (also to physics and to the actuarial sciences) are discussed. In all these cases the attention is focused on open problems and possible extensions of previous results.

The name "conditional functional equations" or "functional equations on restricted domains" is used to denote the class of functional equations for which the set of the values of the variables has to satisfy some additional conditions or restrictions.

Classical examples of conditional equations can be found in information theory because of the probabilistic meaning of the variables: for instance the fundamental

---

AMS (1991) subject classification: 39B52.

Manuscript received January 4, 1993 and, in final form, June 2, 1993.

equation of information

$$f(x) + (1 - x)f\left(\frac{y}{1 - x}\right) = f(y) + (1 - y)f\left(\frac{x}{1 - y}\right), \qquad x, y, x + y \in (0, 1)$$

or the functional equations of sum forms (connected with the characterization of measures of information)

$$\sum_{i=1}^{k} \sum_{j=1}^{l} F_{ij}(p_i, q_j) = 0, \qquad F_{ij} : (0, 1)^2 \to \mathbf{R}$$

where

$$p = (p_1, \ldots, p_k) \in \Gamma_k^\circ := \{p : p_i > 0, \sum p_i = 1\}, \qquad q = (q_1, \ldots, q_l) \in \Gamma_l^\circ.$$

In the last few years several conditional equations were suggested by applications to Economic Theory or, more generally, to the Social and Behaviour Sciences. A typical example is the functional equation for allocation problems (see the (1987) book of J. Aczél on the subject).

As other examples we mention the conditional dilatation equation in normed spaces

$$\|x - y\| = k \implies f(x) - f(y) = g(x, y)(x - y)$$

as well as the numerous Beckman–Quarles type results used to characterize the isometries (W. Benz, J. Lester *et al.*; cf. W. Benz (1992)).

However, the conditional equation which attracted the attention of most of the mathematicians working in this field is the Cauchy equation. Thirty years ago the starting points of these studies were the problems, arising from functional analysis and geometry, suggested respectively by P. Erdös and J. Mikusiński and, some years later, the problem of the extension of local homomorphisms considered by Z. Daróczy and L. Losonczi. These studies generated a very rich research field which is still active and produced more than a hundred research papers.

Here we present some of the results obtained in the last fifteen years (previous results can be found in the excellent survey paper of M. Kuczma (1978)).

Consider the conditional Cauchy equation

$$f(x + y) = f(x) + f(y), \qquad (x, y) \in \Omega \subset G \times G \tag{1}$$

where $f: G \to H$ and $(G, +)$, $(H, +)$ are semigroups, not necessarily commutative. Let

$$\Pi(\Omega) := p_1(\Omega) \cup p_2(\Omega) \cup p_3(\Omega)$$

where $p_i: G \times G \to G$ are the projections defined by

$$p_1(x, y) = x, \qquad p_2(x, y) = y, \qquad p_3(x, y) = x + y.$$

The functional equation (1) constrains the values of f only on $\Pi(\Omega)$, while f can be arbitrary or even undefined on $G \backslash \Pi(\Omega)$. Therefore it is natural to ask the following questions:

(a) Is there a homomorphism g from G into H satisfying $g_{|\Pi(\Omega)} \sim f$ in a suitable sense?
(b) If $\Pi(\Omega) = G$, under which conditions on Ω and f is it possible to assure that $f \in \text{Hom}(G, H)$?

Many results concerning question (a) have been obtained in the past (see the survey paper of M. Kuczma (1978)). Among them the following two are worth a special mention:

— the positive answer given (independently) by S. Kurepa, W. B. Jurkat, and by N. G. de Bruijn to the problem posed by P. Erdös whether an a.e. additive function is a.e. equal to a homomorphism;
— the extension to a homomorphism, in the sense of Z. Daróczy and L. Losonczi, of a function satisfying (1) on a suitable set.

Note that these two results are not comparable since, as shown in the examples by V. Zinde–Walsh, C. T. Ng and J. Aczél (1981), a function may be a.e. additive without having an extension to a homomorphism. The more recent papers by R. Ger (1979) and J. Tabor (1991) answer question (a) in the spirit of the original results of Kurepa–Jurkat–de Bruijn and Daróczy–Losonczi, respectively.

About question (b) some results by L. Paganoni–S. Paganoni Marzegalli (1980) have been obtained under suitable hypotheses on the "bigness" of Ω. Other results by G. L. Forti (1984), C. Borelli–Forti (1987), M. Sablik (1985), W. Jarczyk (1988) and J. Matkowski (1985) establish that f is a homomorphism when Ω is the graph of a function or a special subset of $\mathbf{R}^2$, but in this case some regularity properties of f are required.

Note that (1) implies that the Cauchy equation is satisfied on the set $\tilde{\Omega} \supset \Omega$ defined as the minimal subset of $G \times G$ with the following property:

"whenever three of the points $\{(a, b), (a, b + c), (a + b, c), (b, c)\}$ belong to $\tilde{\Omega}$ then also the fourth does".

So the original problem can be reformulated as follows: under which conditions on Ω, does $\tilde{\Omega} = G \times G$ hold? It seems to us that this aspect has not been completely explored yet. The results in this direction could give, at least in the case $G = \mathbf{R}$ a useful geometric characterization of the sets Ω for which f must be a homomorphism.

Other results are of topological type; for instance the following, obtained by G. L. Forti–L. Paganoni (1990).

Assume G to be a topological group and $\Omega° \neq \varnothing$. If $e \in \Pi(\Omega°)$ and $\{(x, y) \in G \times G: x, y, x + y \in \Pi(\Omega°)\}$ is connected then f is a local homomorphism from $\Pi(\Omega°)$ into H. If, moreover, $\Pi(\Omega°) = X$ then $f \in \mathrm{Hom}(G, H)$.

Among the open problems quoted in the survey paper of Kuczma we mention those concerning the description of the solutions of the following two alternative Cauchy equations:

$$[cf(x + y) - af(x) - bf(y) - d][f(x + y) - f(x) - f(y)] = 0 \tag{2}$$

$$[g(x + y) - g(x) - g(y)][f(x + y) - f(x) - f(y)] = 0. \tag{3}$$

Equation (2) has been completely solved by G. L. Forti (1979) for functions $f: G \to D$, where $(G, +)$ is an abelian group and $(D, \cdot, +)$ an integral domain. He extended previous results by R. Ger (1976) and G. L. Forti–L. Paganoni (1981), and he proved that, except for the case f additive or the special case $a = b = c \neq 0$, $|f(G)| \leq 3$ (here $|X|$ denotes the cardinality of the set X).

It should be interesting to solve equation (2) in the case of more general algebraic structures, for instance when G is a semigroup (not necessarily commutative) and D is a ring.

As to the more difficult equation (3), G. L. Forti–L. Paganoni (1991) have proved, under general hypotheses, that the continuous solutions are trivial, i.e. at least one of the functions f and g has to be a homomorphism. Moreover, they described the class of all solutions when $G = \mathbf{R}^n$, assuming that $\Omega_g := \{(x, y) \in \mathbf{R}^{2n}: g(x + y) \neq g(x) + g(y)\}$ satisfy the property

$$p_1(\Omega_g) = p_1(\Omega_g°) \quad \text{and} \quad p_2(\Omega_g) = p_2(\Omega_g°).$$

Furthermore, in the case $G = \mathbf{R}$, they found that all local solutions of equation (3) in a neighbourhood of the origin are restrictions of global solutions of (3).

It should be interesting to extend these results to general G and to characterize the trivial solutions of (3) under less restrictive regularity properties.

Furthermore we want to mention two other versions of conditional Cauchy equations for which new results have been obtained:

(a) the alternative Cauchy equation $f(x + y) - f(x) - f(y) \in V$ where either V is a suitable set of generators of H or V is a discrete subgroup of H. The description of the general solution has been obtained in the first case by L. Paganoni (1985), G. L. Forti (1987) and C. Borelli Forti–G. L. Forti (1994) and in the second case, under some regularity assumptions, by K. Baron–P. Volkman (1988, 1991), K. Baron–Pl. Kannappan (1990) and M. Sablik (1991).

In the case of an arbitrary set $V \subset H$ the problem of the description of the solutions (even under some regularity assumptions) is still open.

(b) the inhomogeneous Cauchy equation $f(x + y) - f(x) - f(y) = h(x, y)$. For this equation the solutions have been obtained (as a sum of a series) by I. Fenyö (1980), C. Borelli Forti (1989) under the assumption that the n-th Cauchy difference of h is bounded (see also I. Fenyö–D. Rusconi (1981)). In the analytic case other results by I. Fenyö–L. Paganoni (1985) and L. Paganoni–S. Paganoni Marzegalli (1989) gave a characterization of the jacobian elliptic functions. Anyway, except for special results of this type, the problem is still completely open. Any result concerning properties of the solutions depending on the properties of h should be useful to get new information and ideas for all conditional Cauchy equations mentioned before.

Finally we turn to the case when it makes sense to interpret the domain of validity Ω as a (binary) relation of orthogonality. We first consider the conditional Jordan–von Neumann functional equation of orthogonal quadraticity:

$$g: (X, +, \perp) \to (Y, +), \qquad \begin{array}{l} g(x_1 + x_2) + g(x_1 - x_2) = 2g(x_1) + 2g(x_2) \\ \text{for all } x_1, x_2 \in X \text{ with } x_1 \perp x_2. \end{array} \qquad (4)$$

This was first solved by F. Vajzović (1967) in the case where X is a Hilbert space, Y its scalar field, and g is continuous. H. Drljević (1986) replaced ordinary ("euclidean") orthogonality by $x_1 \perp_A x_2 :\leftrightarrow \langle Ax_1, x_2 \rangle = 0$ where A is a selfadjoint operator. M. Fochi (1989) removed the continuity assumption from the former result, and Gy. Szabó (1990) replaced $\perp_A$ by the much more general orthogonality relation $\perp_\varphi$ induced by a sesquilinear functional φ and Y by an abelian group.

The history of orthogonal additivity, i.e., of the conditional Cauchy functional equation

$$f: (X, +, \perp) \to (Y, +), \qquad \begin{array}{l} f(x_1 + x_2) = f(x_1) + f(x_2) \\ \text{for all } x_1, x_2 \in X \text{ with } x_1 \perp x_2. \end{array} \qquad (5)$$

preceded that of (4). It should be noted that there are several concepts of orthogonal additivity. One of them, in which orthogonality in function spaces is given by (μ-almost) disjointness of supports, was initiated by L. Kantorovich–A. Pinsker (1938) and developed further by L. Drewnowski and W. Orlicz and many others; it will not be a subject of this survey. We rather follow the line begun by A. Pinsker (1938) and then furthered by K. Sundaresan (1972), S. Gudder–D. Strawther (1975, 1977) and J. Dhombres (1979). Three of the papers just listed deal with (5) where X is a Hilbert or Banach space, $Y = \mathbf{R}$, and f satisfies some regularity conditions while the 1975 paper introduces a fruitful axiomatization of the orthogonality relation containing the former situations as special cases.

The next step was to get rid of the regularity conditions and to characterize — in an abstract framework — the general even and the general odd solution of (5) with values in an abelian group (J. Lawrence (1985), J. Rätz (1985, 1988, 1989), Gy. Szabó (1986 and more recent papers), J. Rätz–Gy. Szabó (1989). Roughly speaking, in many important situations, the general even solution is quadratic, and the general odd solution is additive. In this way, the range of applications could be broadened essentially.

Among applications inside and outside of mathematics, we mention four:

(a) Where real normed spaces $(X, \| \cdot \|)$ are involved, the so-called Blaschke–Birkhoff–James orthogonality

$$x_1 \perp_{BJ} x_2 :\leftrightarrow \|x_1 + \beta x_2\| \geq \|x_1\| \qquad (\forall \beta \in \mathbf{R})$$

always is in the foreground. It is known (M. M. Day (1947), R. C. James (1947)) that, if dim $X \geq 3$, then $\perp_{BJ}$ is symmetric if and only if X is an inner product space (IPS); this result fails for dim $X = 2$. The 1986 paper by Gy. Szabó gives now a more satisfactory characterization of IPS since it is valid also for dim $X = 2$: If dim $X \geq 2$, $(X, \| \cdot \|)$ is an inner product space if and only if not every $\perp_{BJ}$-additive mapping $f: X \to Y$ is additive.

(b) Among real IPS, it is possible to characterize Hilbert spaces in terms of the boundedness behavior of $\mathbf{R}$-valued orthogonally additive mappings (J. Rätz (1985)). The result is related to the Riesz representation theorem.

(c) An application to physics is the Boltzmann–Gronwall theorem on sum invariants. An application to actuarial science is a premium calculation principle (B. Heijnen–M. J. Goovaerts (1986)).

Many questions in this field need consideration in the future, e.g.: (1) further modification of the general theory in order to include additional examples; (2) building bridges to disciplines such as measure theory, orthogonality theories in special types of algebras, or number-theoretic functions.

Added in proof: Recent progress (J. Rätz (1995), Gy. Szabó (1993, 1995, 1996)) has been achieved in eliminating the homogeneity requirement for the orthogonality relation and in choosing special Z-modules, rather than vector spaces, as domains.

REFERENCES

ACZÉL, J., *A short course on functional equations*. D. Reidel Publ. Co., Dordrecht, 1987.
ACZÉL, J., *Some good and bad characters I have known and where they led. (Harmonic analysis and functional equations)*. [CMS Conf. Proc., Vol. 1]. AMS, Providence, RI, 1981, p. 184.
BARON, K. and KANNAPPAN, PL., *On the Pexider difference*. Fund. Math. *134* (1990), 247–254.
BARON, K. and VOLKMANN, P., *On the Cauchy equation modulo* Z. Fund. Math. *131* (1988), 143–148.
BARON, K. and VOLKMANN, P., *On a theorem of van der Corput*. Abh. Math. Sem. Univ. Hamburg *61* (1991), 189–195.
BENZ, W., *Geometrische Transformationen, unter besonderer Berücksichtigung der Lorentztransformationen*. BI, Mannheim–Leipzig–Wien–Zürich, 1992.
BORELLI FORTI, C., *Condizioni di ridondanza per l'equazione funzionale $f(k(t) + h(t)) = f(K(t)) + f(h(t))$*. Stochastica *11* (1987), 93–105.
BORELLI FORTI, C., *Solutions of a non-homogeneous Cauchy equation*. Radovi Mat. *5* (1989), 213–222.
BORELLI FORTI, C. and FORTI, G. L., *On an alternative functional equation in* R^n. In F. A. N.: *Functional analysis, approximation theory and numerical analysis*. World Scientific Publ. Co., Singapore, Singapore, 1994, pp. 33–44.
DAY, M. M., *Some characterizations of inner-product spaces*. Trans. Amer. Math. Soc. *62* (1947), 320–337.
DHOMBRES, J., *Some aspects of functional equations*. Section 4.9. Chulalongkorn University Press, Bangkok, 1979.
DRLJEVIĆ, F., *On a functional which is quadratic on A-orthogonal vectors*. Publ. Inst. Math. (Beograd) (N.S.) *54* (1986), 63–71.
FENYÖ, I., *Osservazioni su alcuni teoremi di D. H. Hyers*. Istit. Lombardo Accad. Sci. Lett. Rend. A *114* (1980), 235–242.
FENYÖ, I. and RUSCONI, D., *Sulle distribuzioni che soddisfano una equazione funzionale*. Rend. Sem. Mat. Univ. Politec. Torino *39* (1981), 67–76.
FENYÖ, I. and PAGANONI, L., *Su una equazione funzionale proveniente dalla teoria delle funzioni ellittiche jacobiane*. Rend. Mat. Appl. (7) *5* (1985), 319–324.
FENYÖ, I. and PAGANONI, L., *A functional equation which characterizes the jacobian sn(z, k) functions*. Rend. Mat. Appl. (7) *5* (1985), 387–392.
FOCHI, M., *Functional equations in A-orthogonal vectors*. Aequationes Math. *38* (1989), 28–40.
FORTI, G. L., *La soluzione generale dell'equazione funzionale $\{cf(x + y) - af(x) - bf(y) - d\}\{f(x + y) - f(x) - f(y)\} = 0$*. Matematiche (Catania) *34* (1979), 219–242.
FORTI, G. L., *Redundancy conditions for the functional equation $f(x + h(x)) = f(x) + f(h(x))$*. Z. Anal. Anwendungen *3* (1984), 549–554.
FORTI, G. L., *The stability of homomorphisms and amenability, with applications to functional equations*. Abh. Math. Sem. Univ. Hamburg *57* (1987), 215–226.
FORTI, G. L. and PAGANONI, L., *A method for solving a conditional Cauchy equation on abelian groups*. Ann. Mat. Pura Appl. (4) *127* (1981), 79–99.
FORTI, G. L. and PAGANONI, L., *Ω-additive functions on topological groups*. In *Constantin Carathéodory: an international tribute*. World Scientific Publ. Co., Singapore, 1990, 312–330.
FORTI, G. L. and PAGANONI, L., *On an alternative Cauchy equation in two unknown functions. Some classes of solutions*. Aequationes Math. *42* (1991), 271–295.
GER, R., *On a method of solving of conditional Cauchy equations*. Univ. Beograd. Publ. Elektrotechn. Fak. Ser. Mat. Fiz. No. *544–576* (1976), 159–165.
GER, R., *Almost additive functions on semigroups and a functional equation*. Publ. Math. Debrecen *26* (1979), 219–228.

GUDDER, S. and STRAWTHER, D., *Orthogonally additive and orthogonally increasing functions on vector spaces.* Pacific J. Math. *58* (1975), 427–436.

GUDDER, S. and STRAWTHER, D., *A converse of Pythagoras' theorem.* Amer. Math. Monthly *84* (1977), 551–553.

HEIJNEN, B. and GOOVAERTS, M. J., *Additivity and premium calculation principles.* Blätter Deutsch. Ges. Versich. Math. *17* (1986), 217–223.

JAMES, R. C., *Inner products in normed linear spaces.* Bull. Amer. Math. Soc. *53* (1947), 559–566.

JARCZYK, W., *On continuous functions which are additive on their graphs.* [Grazer Ber., No. 292] Forschungsges., Graz, 1988.

KUCZMA, M., *Functional equations on restricted domains.* Aequationes Math. *18* (1978), 1–34.

LAWRENCE, J. *Orthogonality and additive mappings on normed linear spaces.* Colloq. Math. *49* (1985), 253–255.

MATKOWSKI, J., *Cauchy functional equation on a restricted domain and commuting functions.* In *Iteration theory and its functional equations.* Proceedings, Schloss Hofen 1984. [Lecture Notes in Mathematics, No. 1163], Springer Verlag, Berlin, 1985, pp. 101–106.

PAGANONI, L., *On an alternative Cauchy equation.* Aequationes Math. *29* (1985), 214–221.

PAGANONI, L. and PAGANONI MARZEGALLI, S., *Cauchy's functional equation on semigroups.* Fund. Math. *110* (1980), 63–74.

PAGANONI, L. and PAGANONI MARZEGALLI, S., *Holomorphic solutions of an inhomogeneous Cauchy equation.* Aequationes Math. *37* (1989), 179–200.

PINSKER, A., *Sur une fonctionnelle dans l'espace de Hilbert.* C. R. (Doklady) Acad. Sci. URSS N.S. *20* (1938), 411–414.

RÄTZ, J., *On orthogonally additive mappings.* Aequationes Math. *28* (1985), 35–49.

RÄTZ, J., *On orthogonally additive mappings, II.* Publ. Math. Debrecen *35* (1988), 241–249.

RÄTZ, J., *On orthogonally additive mappings, III.* Abh. Math. Sem. Univ. Hamburg *59* (1989), 23–33.

RÄTZ, J., *Orthogonally additive mappings on free product Z-modules.* To appear (1995).

RÄTZ, J. and SZABÓ, GY., *On orthogonally additive mappings, IV.* Aequationes Math. *38* (1989), 73–85.

SABLIK, M., *Note on a Cauchy conditional equation.* Rad. Mat. *1* (1985), 241–245.

SABLIK, M., *A functional congruence revisited.* Aequationes Math. *41* (1991), 273.

SUNDARESAN, K., *Orthogonality and nonlinear functionals on Banach spaces.* Proc. Amer. Math. Soc. *34* (1972), 187–190.

SZABÓ, GY., *On mappings orthogonally additive in the Birkhoff–James sense.* Aequationes Math. *30* (1986), 93–105.

SZABÓ, GY., *Sesquilinear-orthogonally quadratic mappings.* Aequationes Math. *40* (1990), 190–200.

SZABÓ, GY., *On orthogonality spaces admitting nontrivial even orthogonally additive mappings.* Acta Math. Hung. *56* (1990), 177–187.

SZABÓ, GY., *Continuous orthogonality spaces.* Publ. Math. Debrecen *38* (1991), 311–322.

SZABÓ, GY., *Φ-orthogonally additive mappings, I.* Acta Math. Hung. *58* (1991), 101–111.

SZABÓ, GY., *Φ orthogonally additive mappings, II.* Acta Math. Hung. *59* (1992), 1 10.

SZABÓ, GY., *A conditional Cauchy equation on normed spaces.* Publ. Math. Debrecen *42* (1993), 265–271.

SZABÓ, GY., *Isosceles orthogonally additive mappings and inner product spaces.* Publ. Math. Debrecen, to appear (1995).

SZABÓ, GY., *Pyhtagorean orthogonality and additive mappings.* To appear (1996).

TABOR, J., *Cauchy and Jensen equations on a restricted domain almost everywhere.* Publ. Math. Debrecen *39* (1991), 219–235.

VAJZOVIĆ, F., *Über das Funktional H mit der Eigenschaft:* $(x, y) = 0 \Rightarrow H(x + y) + H(x - y) = 2H(x) + 2H(y)$. Glasnik Mat. Ser. III *2 (22)* (1967), 73–81.

Dipartimento di Matematica,
Università di Milano,
Via C. Saldini 50,
I-20133 Milano,
Italia.

Mathematisches Institut,
Universität Bern,
Sidlerstr. 5,
CH-3012 Bern,
Schweiz.

Aequationes Mathematicae **50** (1995) 143–190
University of Waterloo

0001–9054/95/020143–48 $1.50 + 0.20/0
© 1995 Birkhäuser Verlag, Basel

Hyers–Ulam stability of functional equations in several variables

GIAN LUIGI FORTI

Summary. The paper is a survey about Hyers–Ulam stability of functional equations and systems in several variables.
Its content is divided in the following chapters:

1.—Introduction. Historical background.
2.—The additive Cauchy equation; Jensen's equation.
3.—The quadratic equation and the polynomial equation.
4.—The multiplicative Cauchy equation. Superstability.
5.—Approximately multiplicative linear maps in Banach algebras.
6.—Other equations and systems.
7.—Final remarks and open problems.

The bibliography contains 120 items.

1. Introduction. Historical background

Dealing with the stability of functional equations one immediately meets two fields of research which are distinct both in the defining concepts and the methods: functional equations in a single variable and those in several variables.

The present paper is devoted only to equations in several variables and its aim is to present in a more or less organic form the great number of results on the subject published in the last thirteen years. Recently D. H. Hyers and Th. M. Rassias published a survey on stability of homomorphisms ([56]), thus they restricted themselves to the Cauchy functional equation; we intend here to pay attention to a larger class of equations. We make a further restriction by excluding from our considerations the stability of isometries: for problems and results in this field see Hyers's survey [55].

AMS (1991) subject classification: 39B22, 39B32, 39B52

Manuscript received May 12, 1994 and, in final form, September 30, 1994.

As is by now traditional, it is assumed that the following problem, proposed by S. M. Ulam in 1940 (see also [116]), is taken as starting point:

Given a group G and a metric group G' with metric ϱ and given $\varepsilon > 0$, does there exist a $\delta > 0$ such that, if $f: G \to G'$ satisfies the condition

$$\varrho(f(xy), f(x)f(y)) \leq \delta \quad \text{for all } x, y \in G, \tag{1}$$

then a homomorphism $g: G \to G'$ exists such that

$$\varrho(f(x), g(x)) \leq \varepsilon \quad \text{for all } x \in G. \tag{2}$$

In case of a positive answer to the previous problem, we usually say that the homomorphisms from G to G' are **stable** or that the Cauchy functional equation

$$\varphi(xy) = \varphi(x)\varphi(y) \tag{C}$$

is **stable**.

In many problems concerning functional equations related to the Cauchy equation, the main tool is a kind of stability different from that in Ulam's problem. More precisely, it is asked if for every $f: G \to G'$, such that the function

$$(x, y) \mapsto \varrho(f(xy), f(x)f(y))$$

is bounded, there exists a homomorphism $g: G \to G'$ such that the function

$$x \mapsto \varrho(f(x), g(x))$$

is bounded, without paying attention to the bounds and, in particular, without requiring that the second bound goes to zero as the first does. This is the notion of stability most commonly studied and many of the results reported here deal with it.

These two notions of stability are, in general, not equivalent, see the example in Section 7.

The first answer to Ulam's question came within a year when D. H. Hyers [53] proved the following.

THEOREM 1 (D. H. Hyers). *Let B and B' be Banach spaces and let $f: B \to B'$ be a function such that for some $\delta > 0$*

$$\|f(x + y) - f(x) - f(y)\| \leq \delta \quad \text{for all } x, y \in B.$$

Then, for every $x \in B$, the limit $\varphi(x) = \lim\limits_{n \to +\infty} f(2^n x)/2^n$ exists, the function φ is additive and

$$\|f(x) - \varphi(x)\| \leq \delta \qquad \text{for all } x \in B.$$

Moreover φ is the unique additive function satisfying the last inequality.

Furthermore, the continuity of f at a point $y \in B$ implies the continuity of φ on B. The continuity, for each $x \in B$, of the function $t \mapsto f(tx)$, $t \in \mathbb{R}$, implies the homogeneity of φ.

Proof. It is easy to prove, by induction, the inequality

$$\|2^{-n}f(x) - f(2^{-n}x)\| < \delta(1 - 2^{-n}).$$

Put $q_n(x) = f(2^n x)/2^n$, where n is a positive integer and $x \in B$. Then

$$q_n(x) - q_m(x) = \frac{f(2^n x)}{2^n} - \frac{f(2^m x)}{2^m} = \frac{f(2^{m-n} 2^n x) - 2^{m-n} f(2^n x)}{2^m}.$$

Therefore, if $m < n$, we can apply the inequality above and obtain

$$\|q_m(x) - q_n(x)\| < \delta(1 - 2^{m-n})/2^m.$$

Hence $\{q_n(x)\}$ is a Cauchy sequence for each $x \in B$ and, since B' is complete, there exists a limit function $\varphi(x) = \lim\limits_{n \to +\infty} f(2^n x)/2^n$. It is immediate that the function φ is additive and satisfies the inequality $\|f(x) - \varphi(x)\| \leq \delta$ for all $x \in B$.

Suppose that there was another additive function L satisfying the inequality $\|f(x) - L(x)\| \leq \delta$ for all $x \in B$, and such that $L(y) \neq \varphi(y)$ for some $y \in B$. For any integer $n > 2\delta/\|\varphi(y) - L(y)\|$ this means $\|\varphi(ny) - L(ny)\| > 2\delta$, which contradicts the inequalities $\|f(nx) - \varphi(nx)\| \leq \delta$ and $\|f(nx) - L(nx)\| \leq \delta$.

Assume now f continuous at y. If φ is not continuous at a point x, then there exist an integer k and a sequence $\{x_n\}$ of points of B converging to zero such that $\|\varphi(x_n)\| > 1/k$ for all positive integers n. Let m be an integer greater than $3k\delta$. Then $\|\varphi(mx_n + y) - \varphi(y)\| = \|\varphi(mx_n)\| > 3\delta$. On the other hand,

$$\|\varphi(mx_n + y) - \varphi(y)\|$$
$$\leq \|\varphi(mx_n + y) - f(mx_n + y)\| + \|f(mx_n + y) - f(y)\| + \|f(y) - \varphi(y)\| < 3\delta$$

for sufficiently large n, since $\lim\limits_{n \to +\infty} f(mx_n + y) = f(y)$. This contradiction establishes the theorem.

For fixed x, $f(tx)$ is continuous in t; thus $\varphi(tx)$ is additive and continuous in t, hence it is homogeneous.   □

Thus, when G and G' are the additive groups of Banach spaces, we have a positive answer to Ulam's question with $\varepsilon = \delta$; thus in this case the Cauchy equation is stable or the homomorphisms are stable. (A similar result was proved in 1924 by G. Pólya and G. Szegö for $G = \mathbb{N}$ and $G' = \mathbb{R}$; see [72].)

If we carefully look at the proof of the main part of Hyers's theorem, the existence of the additive function φ uniformly approximating f, we easily recognize that the result is true if we replace the additive group of the Banach space B by a commutative semigroup S. So we can conclude that the homomorphisms from an abelian semigroup into the additive group of a Banach space are stable.

It is natural to generalize the notion of stability to other functional equations and to other possible kinds of approximations. Thus, if we are in an appropriate framework and have the functional equation

$$E_1(\varphi) = E_2(\varphi) \qquad\qquad (E)$$

and a function f which is an *approximate* solution of (E), i.e., $E_1(f)$ and $E_2(f)$ are *close* in some sense, we may ask whether a solution φ of (E) exists *near f*.

After Hyers's result a great number of papers on the subject have been published, generalizing Ulam's problem and Hyers's theorem in various directions. The exposition of these results constitute the content of the following sections.

The division into different sections is somehow artificial; for instance results about the additive Cauchy equation and the polynomial equation are often consequences of the same theorem; nevertheless the author hopes to achieve in this way a greater clarity in the exposition.

To finish this introductory section we mention some studies of J. Baster, Z. Moszner and J. Tabor ([7], [67], [68]) concerning different notions of stability and their mutual relations.

In particular they point out that, when considering the functional equation (E), usually it is possible to write it in equivalent forms, say

$$E_1'(\varphi) = E_2'(\varphi)$$

where $E_1' \neq E_1$ and $E_2' \neq E_2$ (for example the Cauchy equation can be written as $f(xy) - f(x) - f(y) = 0$ or $f(xy) = f(x) + f(y)$) and the use of different forms of the same functional equation is not indifferent to stability: to measure the *distance* from $E_1(f)$ to $E_2(f)$ or that from $E_1'(f)$ to $E_2'(f)$ can lead (also for the Cauchy equation) to completely different conclusions.

In the last section of the present paper we shall return to this and similar problems.

2. The additive Cauchy equation; Jensen's equation

This field is by far the richest in results and, to some extent, the best organized. We consider the additive Cauchy equation

$$\varphi(xy) = \varphi(x) + \varphi(y); \tag{A}$$

obviously this equation is exactly the same as equation (C), but with the use of the additive notation on the right-hand side we emphasize that the range of the function is a vector space of which we consider the additive group.

In the first part of this section we treat various generalizations of Hyers's theorem from the point of view of the domain of the functions involved.

We start with the following.

DEFINITION 1. Let G be a semigroup and B a Banach space. We say that equation (A) is *stable for the pair* (G, B) if, for every function $f: G \to B$ such that

$$\|f(xy) - f(x) - f(y)\| \le \delta, \qquad x, y \in G \qquad \text{for some } \delta \ge 0, \tag{3}$$

there exists a solution φ of (A) such that

$$\|f(x) - \varphi(x)\| \le \varepsilon, \qquad x \in G \tag{4}$$

for some ε depending only on δ.

We point out that the notion of stability here defined is, at least apparently, different than that contained in the original Ulam's problem: we simply require that ε depend on δ, not that for each $\varepsilon > 0$ there exists a $\delta > 0$ such that from (3) we get (4).

It has been proved (see for instance [35]) that in this context the only property of B involved is the completeness; more precisely if B_1, B_2 are Banach spaces, then (A) is stable for (G, B_1) if and only if it is for (G, B_2). Moreover, when (A) is stable, the smallest bound ε in (4) is equal to δ and the additive function φ satisfying (4) is unique (see [34]); so *a posteriori* Definition 1 coincides with the notion of stability in Ulam's problem.

Due to the previous remark we simply say that (A) is stable for a group or a semigroup G; moreover the stability of (A) simply means that any function $f: G \to \mathbb{R}$ satisfying (3) can be decomposed as sum of an additive function φ and a bounded function b:

$$f(x) = \varphi(x) + b(x), \qquad x \in G.$$

Thus Theorem 1 says that (A) is stable for any commutative semigroup G.

Before looking for other classes of groups and semigroups for which (A) is stable, we note that, for a function f satisfying (3), the limit of Hyers's sequence $\{2^{-n}f(x^{2n})\}$ exists for all x whatever G is; the point is that the function φ so obtained satisfies the functional equation $\varphi(x^2) = 2\varphi(x)$, but in general it need not be additive.

A very simple result shows that the additive Cauchy equation is stable for any group having only elements of finite order and in this case the completeness of the range is not necessary (see [34]).

A remarkable achievement was that of L. Székelyhidi who in [102] replaced the original proof given by Hyers by a new one based on the use of invariant means. We remind here that a semigroup G is said left (right) amenable if there exists a left (right) invariant mean on the space of the bounded complex functions defined on G (for the definition and properties of invariant means and amenable semigroups see [51]).

THEOREM 2 (L. Székelyhidi). *Let G be a left (right) amenable semigroup, then* (A) *is stable for G.*

Proof. By the previous observations it is enough to consider complex functions. So let $f: G \to \mathbb{C}$ be such that $|f(xy) - f(x) - f(y)| \le \delta$ for all $x, y \in G$ and some $\delta \ge 0$; then for each fixed $x \in G$, the function $y \mapsto f(xy) - f(y)$ is bounded. Let m_y be a left invariant mean on the space of bounded complex functions defined on G (the suffix y denotes that m_y acts on functions of the variable y) and define

$$\varphi(x) = m_y\{_xf - f\}, \qquad x \in G,$$

where $_xf(t) = f(xt)$. We have

$$\varphi(xz) = m_y\{_{xz}f - f\} = m_y\{_{xz}f - _xf + _xf - f\} = m_y\{_x(_zf - f)\} + m_y\{_xf - f\}$$
$$- m_y\{_zf - f\} + m_y\{_xf - f\} - \psi(z) + \psi(x),$$

so φ is a solution of (A). Moreover we have

$$|\varphi(x) - f(x)| = |m_y\{_xf - f\} - f(x)| = |m_y\{_xf - f - f(x)\}|$$
$$\le \sup_{y \in G} |f(xy) - f(x) - f(y)| \le \delta,$$

for all $x \in G$; thus (A) is stable on G. $\qquad\square$

Now a question naturally arises: do groups or semigroups exist for which equation (A) is not stable? In view of Székelyhidi's theorem we must look among non-amenable groups or semigroups and in fact the present author in [34] proved that on the free group (or the free semigroup), generated by 2 elements, the additive Cauchy equation (A) is not stable. These results suggested the study of connections between stability and amenability.

It is well known that any subgroup of an amenable group is amenable, but this property is not true of stability. J. Lawrence proved that **any** torsion-free group can be embedded in a group G for which (A) is stable in a trivial sense, namely any function on G satisfying (3) is bounded and so $\varphi = 0$ (see [36]).

Thus stability is strictly weaker than amenability; nevertheless there is an equivalence between amenability and a kind of multi-stability (see [34]).

Recently L. Giudici proved the following (still unpublished) result.

LEMMA 1 (L. Giudici). *Let G be a group. The Cauchy equation (A) is stable for G if and only if each function $f: G \to \mathbb{R}$ satisfying (3) is bounded on the commutator subgroup G^1 of G.*

(For the definition of G^1 see for instance [85].)

Proof. Since any additive function is obviously identically zero on G^1, the "only if" part is proved.

Now assume that f satisfies (3) and $|f(x)| \le K$ on G^1 with some $K \ge 0$. Let $\{x_\alpha\}$ be a complete system of representatives of the cosets of G^1 in G and define $\tilde{f}: G/G^1 \to \mathbb{R}$ as $\tilde{f}(x_\alpha G^1) := f(x_\alpha)$. We show that $\tilde{f}$ has bounded Cauchy difference. Take x_α, x_β and let $x_\gamma \equiv \chi_\alpha x_\beta \pmod{G^1}$:

$$\left| \tilde{f}(x_\alpha G^1 x_\beta G^1) - \tilde{f}(x_\alpha G^1) - \tilde{f}(x_\beta G^1) \right|$$

$$= \left| \tilde{f}(x_\gamma G^1) - \tilde{f}(x_\alpha G^1) - \tilde{f}(x_\beta G^1) \right|$$

$$\le \left| f(x_\alpha x_\beta) - f(x_\alpha) - f(x_\beta) \right| + \left| f(x_\alpha x_\beta) - f(x_\gamma) - f(x_\alpha x_\beta x_\gamma^{-1}) \right| + \left| f(x_\alpha x_\beta x_\gamma^{-1}) \right|$$

$$\le 2\delta + K.$$

Since G/G^1 is abelian, (A) is stable on it and so

$$\tilde{f} = \tilde{\varphi} + \tilde{b}$$

where $\tilde{\varphi}: G/G^1 \to \mathbb{R}$ is additive and $\tilde{b}$ is bounded.

If we define $\varphi: G \to \mathbb{R}$ by $\varphi(x) := \tilde{\varphi}(xG^1)$, then φ is additive. To finish we must prove that $b(x) := f(x) - \varphi(x)$ is bounded. Take $x \in G$ and let $x_\alpha \equiv x \pmod{G^1}$;

then

$$
\begin{aligned}
\left|b(x)\right| = \left|f(x) - \varphi(x)\right| &= \left|f(x) - f(x_\alpha) - f(xx_\alpha^{-1}) + f(xx_\alpha^{-1}) + f(x_\alpha) - \tilde{\varphi}(x_\alpha G^1)\right| \\
&\leq \left|f(x) - f(x_\alpha) - f(xx_\alpha^{-1})\right| + \left|f(xx_\alpha^{-1})\right| + \left|\tilde{b}(x_\alpha G^1)\right| \\
&\leq \delta + K + \left|\tilde{b}(x_\alpha G^1)\right| \\
&\leq K'
\end{aligned}
$$

since $\tilde{b}$ is bounded.   □

From Lemma 1 we easily obtain the following.

THEOREM 3 (L. Giudici). *Let G be a group and assume there exists an integer N such that each element of the commutator subgroup G^1 is the product of at most N commutators. Then (A) is stable on G.*

Proof. Assume $f: G \to \mathbb{R}$ is such that

$$
\left|f(xy) - f(x) - f(y)\right| \leq \delta, \qquad x, y \in G.
$$

Let $u \in G^1$ with $u = t_1 t_2 \ldots t_s$, where $s \leq N$ and the t_i's are commutators. First we show that f is bounded on the commutators:

$$
\begin{aligned}
\left|f(x^{-1}y^{-1}xy)\right| &\leq \left|f(x^{-1}y^{-1}xy) + f(yx) - f(xy)\right| + \left|f(xy) - f(x) - f(y)\right| \\
&\quad + \left|f(x) + f(y) - f(yx)\right| \\
&\leq 3\delta.
\end{aligned}
$$

Thus from

$$
\begin{aligned}
f(u) = f(t_1 t_2 \ldots t_s) &= \{f(t_1 t_2 \ldots t_s) - f(t_1) - f(t_2 \ldots t_s)\} \\
&\quad + \{f(t_2 \ldots t_s) - f(t_2) - f(t_3 \ldots t_s)\} + \cdots \\
&\quad + f(t_1) + \cdots + f(t_s)
\end{aligned}
$$

we obtain $\left|f(u)\right| \leq 4N\delta$.

The theorem follows from Lemma 1.   □

Since, for example, the group $GL_2(\mathbb{R})$ is not amenable but has the property required by Theorem 3 with $N = 2$, we actually obtain a new class of groups for which (A) is stable.

From the results of J. Rätz contained in [84] we get another class of groups (and semigroups) for which (A) is stable

THEOREM 4 (J. Rätz). *Let G be a semigroup such that for each pair $x, y \in G$ there exists an integer $n(x, y) \geq 2$ such that*

$$(xy)^{n(x,y)} = x^{n(x,y)}y^{n(x,y)}.$$

Then equation (A) is stable on G.

It is possible to show that the three classes of groups considered in Theorems 2, 3 and 4 are distinct, i.e., no one contains another.

What happens if in Definition 1 we do not require the completeness of B? This property is "almost" necessary for stability; more precisely J. Schwaiger and the present author proved in [37] the following.

THEOREM 5 (G. L. Forti and J. Schwaiger). *Let G be an abelian group with an element of infinite order. Assume that equation (A) is stable in the sense of Definition 1, where B is a normed space. Then B is complete.*

The notion of stability given in Definition 1 can be generalized in various ways for functions taking values in vector spaces which are not necessarily normed. For example it is enough to have some notion of boundedness. More generally what is important is a connection between the range of the Cauchy difference $f(xy) - f(x) - f(y)$ and that of $f(x) - \varphi(x)$.

The first very general and interesting results in this direction have been proved by J. Rätz in [84]. Here, just to give the flavour of the paper, we state a particular case of the main theorem.

THEOREM 6 (J. Rätz). *Let G be a semigroup satisfying the condition of Theorem 4. Let Y be a $\mathbb{Q}$-topological vector space sequentially complete and let V be a non-empty, $\mathbb{Q}$-convex and bounded subset of Y containing 0. Then for any function $f: G \to Y$ such that*

$$f(xy) - f(x) - f(y) \in V, \qquad x, y \in G,$$

there exists a solution φ of (A) such that

$$f(x) - \varphi(x) \in \text{seqcl}(V), \qquad x \in G.$$

($\text{seqcl}(V)$ is the sequential closure of V).
Moreover, if Y is Hausdorff then φ is unique.

By using Rätz's result, Z. Gajda in [39] proved the following generalization of a result of L. Székelyhidi ([103]).

THEOREM 7 (Z. Gajda). *Let G be a semigroup for which equation (A) is stable for complex functions. Let Y be a Hausdorff topological vector space, locally convex and sequentially complete.*
  For any function f: G → Y such that the function

$$(x, y) \mapsto f(xy) - f(x) - f(y)$$

is bounded, there exists a solution φ of (A) such that f − φ is bounded.

K. Nikodem in [70] studied the stability of the Pexider equation

$$\varphi_1(xy) = \varphi_2(x) + \varphi_3(y) \tag{P}$$

and, by using the results of Z. Gajda and R. Ger in [45], proved the following.

THEOREM 8 (K. Nikodem). *Let G be a commutative semigroup. Let Y be a ℚ-topological vector space sequentially complete and let V be a non-empty, ℚ-convex, symmetric and bounded subset of Y. If $f_i: G → Y$, $i = 1, 2, 3$, satisfy the condition*

$$f_1(xy) - f_2(x) - f_3(y) \in V, \qquad x, y \in G,$$

then there exists a triplet $(\varphi_1, \varphi_2, \varphi_3)$ solution of equation (P) such that

$$f_1(x) - \varphi_1(x) \in 3\mathrm{seqcl}(V), \qquad f_2(x) - \varphi_2(x) \in 4\mathrm{seqcl}(V),$$
$$f_3(x) - \varphi_3(x) \in 4\mathrm{seqcl}(V).$$

A stability result for the Pexider equation for set valued mappings has been proved by A. Smajdor in [97].

It should be noted that, when the space of the values is not normed, a theorem of completeness like Theorem 5 doesn't hold. Indeed L. Giudici proved the existence of a locally convex Hausdorff topological vector space Y, not sequentially complete, such that we have stability of (A) for the functions from ℝ into Y.
  The definition 1 of stability is obviously based on Hyers's result, so it concerns functions having values in a Banach space. All other results presented till now, as generalizations of Hyers's, are about functions with values in a vector space. The original problem of Ulam was about metric groups which are not necessarily the additive groups of some vector space.

Studying the stability problem in this more general framework, D. Cenzer in [14] considered transformations of the circle. Namely, let G be the group represented by the interval $[0, 1)$ with addition modulo 1 and define a metric ϱ in G as follows:

$$\varrho(x, y) = \min\{|x - y|, |1 - x + y|\};$$

with this metric G becomes a complete metric group. If $f: G \to G$ satisfies (1) with $\delta < 1/6$, then there exists a homomorphism $\varphi: G \to G$ satisfying (2) with $\varepsilon = \delta$. This result is easily extended to the k-dimensional torus and to the dyadic rationals.

About this general form of the stability problem very few results are known and they are negative: L. Paganoni in [71] noted that if $f: \mathbb{Z} \to \mathbb{Z}$ is defined as $f(x) = [x/2]$, then $|f(x + y) - f(x) - f(y)| \le 1$ but for every additive $\varphi: \mathbb{Z} \to \mathbb{Z}$ the difference $f - \varphi$ is unbounded. Analogous results were proved by D. Kazhdan in [62].

The second part of this section is devoted to a different generalization of the original Ulam problem.

We consider functions defined on a normed space X and taking values in a Banach space Y; thus we shall use additive notation.

We assume that $f: X \to Y$ satisfies an inequality of the form

$$\|f(x + y) - f(x) - f(y)\| \le \Phi(x, y), \tag{5}$$

where $\Phi: X \times X \to \mathbb{R}$ is a given function, usually satisfying some growing condition; sometimes in what follows the function Φ will not be defined on the whole space $X \times X$ but only on $(X \times X)\backslash(0, 0)$: this will be clear from the context. We ask for the existence of an additive φ such that

$$\|f(x) - \varphi\| \le \Psi(x), \tag{6}$$

where Ψ is a function we can explicitly compute starting from Φ.

It seems that D. G. Bourgin in 1951 has been the first author treating this problem; he simply stated (without proof) in [12] that, if Φ depends on $\|x\|$ and $\|y\|$, is monotonic, nondecreasing and symmetric in $\|x\|$ and $\|y\|$ and moreover the series

$$\sum_{i=1}^{+\infty} 2^{-i} \Phi(2^{i-1}x, 2^{i-1}x) \tag{7}$$

converges for each $x \in X$, then (6) holds with Ψ equal to the sum of the series (7).

In 1978 Th. M. Rassias in [78] proved a stability theorem where

$$\Phi(x, y) = \theta(\|x\|^p + \|y\|^p), \qquad p \in [0, 1), \qquad \theta \ge 0,$$

and he got a unique additive φ satisfying (6) with

$$\Psi(x) = \frac{2\theta}{2 - 2^p} \|x\|^p,$$

i.e., a special case of Bourgin's result. Rassias moreover required the continuity of $f(tx)$ for each fixed x, as a function of the real variable t; from this additional hypothesis he got not only the additivity but also the linearity of φ.

Many other similar results have been published:

—J. M. Rassias in [73] and [74] with $\Phi(x, y) = \theta \|x\|^p \|y\|^p, \theta \geq 0, p \in [0, 1/2)$, obtaining

$$\Psi(x) = \frac{\theta}{2 - 2^p} \|x\|^{2p};$$

—J. M. Rassias in [75] and [76] with $\Phi(x, y) = \theta \|x\|^p \|y\|^q, \theta \geq 0, p + q \in [0, 1)$,

$$\Psi(x) = \frac{\theta}{2 - 2^{p+q}} \|x\|^{p+q};$$

—G. Isac and Th. M. Rassias in [57] with $\Phi(x, y) = \theta(\psi(\|x\|) + \psi(\|y\|))$, where $\theta \geq 0$ and $\psi : \mathbb{R}^+ \to \mathbb{R}^+$ is such that $\psi(t)/t \to 0$ as $t \to +\infty$, $\psi(ts) \leq \psi(t)\psi(s)$ and $\psi(t) < t$ for $t > 1$; the function Ψ becomes

$$\Psi(x) = \frac{2\theta}{2 - \psi(2)} \psi(\|x\|);$$

—Th. M. Rassias and P. Šemrl in [83] with $\Phi(x, y) = H(\|x\|, \|y\|)$, where $H : \mathbb{R}^+ \times \mathbb{R}^+ \to \mathbb{R}^+$ is symmetric, increasing in each variable and homogeneous of degree $p \in [p, +\infty) \setminus \{1\}$, obtaining

$$\Psi(x) = \frac{H(1, 1)}{|2 - 2^p|} \|x\|^p.$$

In all these papers an additional hypothesis of continuity, similar to that indicated above, suffices to prove that the function φ is linear.

It should be noted that all such results (the last only for $p < 1$) concerning the existence of the additive φ and the evaluation of the approximation are particular cases of a stability theorem for a class of functional equations of the form

$g[F(x, y)] = H[g(x), g(y)]$ (thus containing the Cauchy equation) proved in [33]: there it is simply required that the series (7) converges and $2^{-i}\Phi(2^{i-1}x, 2^{i-1}y) \to 0$ as $i \to \infty$, for every x, y.

In [41] Z. Gajda extended the result proved in [78] by Th. M. Rassias to the case $p > 1$ and noted that it holds for $p < 0$ as well. Moreover he proved with an example (see also [61] and [83]) that the case $p = 1$ is critical. More precisely there exists a function $f: \mathbb{R} \to \mathbb{R}$ satisfying

$$|f(x + y) - f(x) - f(y)| \leq \theta(|x| + |y|)$$

for some positive θ and such that for any additive φ the quotient

$$\frac{|f(x) - \varphi(x)|}{|x|}, \qquad x \neq 0,$$

is unbounded.

It is natural to ask whether by a slight diminishing of the right-hand side in the previous inequality one might get stability. This question recently led R. Ger to study the inequality (5) with some other control functions Φ. First he obtained a stability result in finite dimensional normed spaces ([48]), later in a more general framework. We state here, for sake of simplicity, only a particular case of the results proved in [50].

THEOREM 9 (R. Ger). *Let $(G, +)$ be an amenable group and $(Y, \|.\|)$ a real reflexive Banach space. Let $f: G \to Y$ be a function satisfying the inequality* (5) *where Φ has one of the following forms:*
    (i) $\Phi(x, y) = F(x)$;
    (ii) $\Phi(x, y) = F(x) + F(y) - F(x + y)$, *F even and subadditive*;
    (iii) $\Phi(x, y) = F(x + y)$, $F(0) = 0$;
where F is a given function from G into $[0, \infty)$.
    Then there exists an additive $\varphi: G \to Y$ such that

$$\|f(x) - \varphi(x)\| \leq F(x), \qquad x \in G$$

in cases (i) *and* (iii), *or*

$$\|f(x) - \varphi(x)\| \leq 2F(x), \qquad x \in G$$

in case (ii).
    Moreover, if $\lim_{k \to +\infty} F(kx)/k = 0$ for all $x \in G$, then φ is uniquely determined.

The proof of Theorem 9 makes use of some results of Z. Gajda ([44]) and R. Badora ([4]) extending the notion of amenability to vector-valued mappings. The case (i) has also been treated by B. E. Johnson in [61].

An analogue of Theorem 9 has been proved by Z. Gajda in [43] under the same hypotheses on G and Y and by assuming $\Phi(x, y) = F_1(x) + F_2(y)$ with

$$\inf\left\{\sum_{i=1}^{k} F_1(y_i x): x \in G\right\} = 0 \qquad \text{for any } k \in \mathbb{Z} \text{ and any } y_1, y_2, \ldots, y_k \in G.$$

Then condition (6) is satisfied with $\Psi = F_2$.

Another possibility is to consider the inequality

$$\left|f\left(\sum_{i=1}^{n} x_i\right) - \sum_{i=1}^{n} f(x_i)\right| \leq \theta \sum_{i=1}^{n} |x_i|, \qquad f: \mathbb{R} \to \mathbb{R},$$

and assume its validity for all $n \in N$, all $x_1, \ldots, x_n \in \mathbb{R}$ and some $\theta > 0$; P. Šemrl in [86] proved that in this case if f is continuous then there exists an additive $\varphi: \mathbb{R} \to \mathbb{R}$ such that $|f(x) - \varphi(x)| \leq \theta|x|$.

J. Tabor in [113] and [114] considered the case when Φ depends on the function f and got some regularity properties of f itself. J. Chmielinski and J. Tabor in [15] studied the analogous problem for the Pexider equation.

To finish this section on the Cauchy functional equation we report some results concerning inequalities (3) or (5) on **restricted domains**.

In order to present them we recall the notion of proper linearly invariant ideal and σ-ideal of sets and that of conjugate ideals of sets.

DEFINITION 2. Let $(G, +)$ be an abelian group. A non-empty family $\mathscr{F}$ of subsets of G is called *proper linearly invariant ideal* (in short p.l.i. ideal) if the following conditions are fulfilled:

(i) $A, B \in \mathscr{F}$ imply $A \cup B \in \mathscr{F}$;
(ii) $A \in \mathscr{F}$ and $B \subset A$ imply $B \in \mathscr{F}$;
(iii) $x \in G$ and $A \in \mathscr{F}$ imply $x - A \in \mathscr{F}$;
(iv) $G \notin \mathscr{F}$.

If (i) is replaced by

(v) $\mathscr{F}$ is closed under countable unions;

then $\mathscr{F}$ will be called a p.l.i. σ-ideal.

We say that two p.l.i. ideals $\mathscr{F}_1$ and $\mathscr{F}_2$ in G and $G \times G$, respectively, are *conjugate* whenever for every set $M \in \mathscr{F}_2$ there exists a set $U \in \mathscr{F}_1$ such that $M_x = \{y \in G: (x, y) \in M\} \in \mathscr{F}_1$ provided $x \in G \setminus U$.

We can now state the result of R. Ger ([46]) which may be considered a restricted domain version of Theorem 6.

THEOREM 10 (R. Ger). *Let $(G, +)$ be an abelian group uniquely 2-divisible and let $\mathscr{F}_1$ and $\mathscr{F}_2$ be conjugate p.l.i. σ-ideals in G and $G \times G$ respectively. Assume that*
   (i) $A \in \mathscr{F}_1$ *implies* $\frac{1}{2}A \in \mathscr{F}_1$;
   (ii) $A \in \mathscr{F}_1$ *implies* $(A \times G) \cup (G \times A) \in \mathscr{F}_2$;
   (iii) $M \in \mathscr{F}_2$ *implies* $T(M) \in \mathscr{F}_2$ *where* $T: G \times G \to G \times G$ *is given by*

$$T(x, y) = (-x + y, x).$$

Let Y be a $\mathbb{Q}$-topological vector space sequentially complete and let V be a non-empty, $\mathbb{Q}$-convex, symmetric and bounded subset of Y containing 0. If $f: G \to Y$ satisfies the condition

$$f(x + y) - f(x) - f(y) \in V, \qquad (x, y) \in G^2 \backslash M, \tag{8}$$

where $M \in \mathscr{F}_2$, then there exist an additive function $\varphi: G \to Y$ and a set $E \in \mathscr{F}_1$ such that

$$f(x) - \varphi(x) \in 3seqcl(V), \qquad x \in G \backslash E. \tag{9}$$

If Y is a T_1 topological space, then φ is unique.

In [111] J. Tabor proved a similar result without assuming G to be abelian. If the following conditions are fulfilled:

   (i) there is a subsemigroup S of G, $S \notin \mathscr{F}_1$, such that $S - S = G$;
   (ii) for each pair $x, y \in S$, there exists $n \in \mathbb{N}$ such that $2^n(x + y) = 2^n x + 2^n y$;
   (iii) for any $A \in \mathscr{F}_1$, the set $\{x \in G: \exists n \in \mathbb{N} \cup \{0\}$ such that $2^n x \in A\}$ is in $\mathscr{F}_1$;

and (8) is satisfied on $S^2 \backslash M$, then (9) is satisfied on $S \backslash E$. An analogous result holds for the Pexider equation (P), with $S = G$. (See also [112].)

Following this line of research, Z. Gajda in [42], [43] and [44] introduced the space $L_{\mathscr{F}}^\infty(F, \mathbb{R})$ of the real functions defined on a group G and $\mathscr{F}$-essentially bounded, where $\mathscr{F}$ is a p.l.i. ideal, i.e.,

$$f \in L_{\mathscr{F}}^\infty(G, \mathbb{R}) \quad \text{if and only if there exists } A \in \mathscr{F} \text{ such that } \sup_{x \in G \backslash A} |f(x)| < +\infty.$$

Gajda generalized the notion of invariant mean to the space $L_{\mathscr{F}}^\infty(G, \mathbb{R})$ and then to the case of vector-valued functions. After giving some conditions concerning the existence of such a mean, he was able to prove theorems of stability on restricted domains. An example is the following.

THEOREM 11 (Z. Gajda [42], [44]). *Let* $(G, +)$ *be a group,* $\mathscr{F}_1$ *and* $\mathscr{F}_2$ *be conjugate p.l.i. ideals in G and* $G \times G$ *respectively. Let Y be a Hausdorff topological vector space and assume there exists an invariant mean on* $L_{\mathscr{F}}^{\infty}(G, Y)$. *Let V be a bounded subset of Y and let* $f: G \to Y$ *be a mapping satisfying* (8). *Then there exists exactly one additive* $\varphi: G \to y$ *such that*

$$f(x) - \varphi(x) \in \operatorname{convcl}(V), \qquad x \in G \backslash E,$$

for some $E \in \mathscr{F}_1$.

F. Skof in [93] studied a more traditional form of local stability. She considered a function f from a subset D of the reals into a Banach space Y satisfying condition (3) for all $(x, y) \in E \subset \mathbb{R}^2$. The following five cases were treated:

(1)— $E = \mathbb{R}^+ \times \mathbb{R}^+, D = \mathbb{R}^+$;
(2)— $E = \{(x, y) \in \mathbb{R}^2 : 0 \leq x < a, 0 \leq y < a, 0 \leq x + y < a\}, D = [0, a)$;
(3)— $E = \{(x, y) \in \mathbb{R}^2 : x^2 + y^2 < a\}, D = (-a\sqrt{2}, a\sqrt{2})$;
(4)— $E = \{(x, y) \in \mathbb{R}^2 : (x - \alpha)^2 + (y - \beta)^2 < a^2\}$,
        $D = (\alpha - a, \alpha + a) \cup (\beta - a, \beta + a) \cup (\alpha + \beta - a, \alpha + \beta + a)$,
        for some $\alpha, \beta \in \mathbb{R}$;
(5)— $E = \{(x, y) \in \mathbb{R}^2 : |x| + |y| > a\}, D = \mathbb{R}$;

where a is a positive number.

By extending the function f defined on D to a function $\hat{f}$ defined on the whole $\mathbb{R}$ still satisfying (3) with a possibly different δ, and then by using Hyers's theorem, she got an additive $\varphi: \mathbb{R} \to Y$ such that

$$\|f(x) - \varphi(x)\| \leq \mu, \qquad x \in D,$$

where, in the five cases above, μ equals δ, 3δ, 11δ, $34\delta + \|f(\alpha) - \varphi(\alpha)\| + \|f(\beta) - \varphi(\beta)\|$ and 9δ, respectively. (Note that in this case of local stability, the bound μ does not only depend on δ but also on the functions f and φ.)

By using Skof's extension theorems, Z. Kominek in [63] proved analogous results in $\mathbb{R}^n$. Moreover he obtained the following theorem on Jensen's equation.

THEOREM 12 (Z. Kominek). *Let* $D \subset \mathbb{R}^n$ *be a bounded set with non-empty interior and assume there exists* $x_0 \in D^\circ$ *such that* $D_0 = D - x_0$ *satisfies the condition* $1/2 D_0 \subset D_0$. *For each function* $f: D \to Y$ *fulfilling*

$$\left\| f\left(\frac{x + y}{2}\right) - \frac{f(x) + f(y)}{2} \right\| \leq \delta, \qquad x, y \in D,$$

for some δ ≥ 0, there exist g: $\mathbb{R}^n \to Y$ and K > 0 such that

$$g\left(\frac{x+y}{2}\right) = \frac{g(x) + g(y)}{2} \qquad \text{for } x, y \in \mathbb{R}^n \text{ and } \|f(x) - g(x)\| \leq K \text{ for } x \in D.$$

A classical problem when studying the Cauchy equation on restricted domains is that of the orthogonally additive functions, i.e., functions which are assumed to satisfy (A) for orthogonal pairs (x, y), obviously in a setting where some kind of orthogonality is defined.

The analogous stability problem has been attacked by H. Drljević in [25], [26], [27] and by H. Drljević and Z. Mavar in [28], in different classes of spaces. In [28] the following is proved (see also [27]).

THEOREM 13 (H. Drljević and Z. Mavar). *Let* $f: X \to \mathbb{C}$ *where X is a complex Hilbert space and assume that for every fixed* $x \in X$ *the function* $t \mapsto f(tx)$, $t \in \mathbb{C}$, *is continuous.*

Let $T: X \to X$ *be a continuous selfadjoint operator with* dim $T(X) = 1$ *or* 2. *If*

$$|f(x+y) - f(x) - f(y)| \leq \theta[|(Tx, x)|^p + |(Ty, y)|^p]$$

for some $\theta \geq 0$, $p \in [0, 1/2)$ *and for all* $x, y \in X$ *such that* $(Tx, y) = 0$, *then there exists a unique continuous* $\varphi: X \to \mathbb{C}$, *additive on the pairs* (x, y) *with* $(Tx, y) = 0$ *and such that*

$$|f(x) - \varphi(x)| \leq \theta_1 |(Tx, x)|^p, \qquad x \in X,$$

for some constant θ_1 $((.,.)$ *denotes the inner product in X).*

3. The quadratic equation and the polynomial equation

In the first part of this section we consider the quadratic equation

$$\kappa(x + y) + \kappa(x - y) = 2\kappa(x) + 2\kappa(y), \qquad (Q)$$

where $\kappa: G \to Y$, G group and Y vector space.

The natural extension of Ulam's problem to the equation (Q) is the following: we assume Y to be normed space and that $f: G \to Y$ fulfils the inequality

$$\|f(x + y) + f(x - y) - 2f(x) - 2f(y)\| \leq \delta, \qquad x, y \in G, \qquad (10)$$

for some $\delta \geq 0$; hence we look for a solution $\kappa\colon G \to Y$ of (Q) satisfying

$$\|f(x) - \kappa(x)\| \leq \varepsilon, \qquad x \in G, \tag{11}$$

for some ε depending only on δ.

The first author treating the problem above was F. Skof, who in [94] proved the following.

THEOREM 14 (F. Skof). *Let X be a normed vector space and Y a Banach space. If $f\colon X \to Y$ fulfils (10), then for every $x \in X$ the limit*

$$\kappa(x) = \lim_{n \to +\infty} \frac{f(2^n x)}{2^{2n}}$$

exists and κ is the unique solution of (Q) satisfying (11) with $\varepsilon = \delta/2$.

Notice that Theorem 14 is of a form very similar to that of Hyers's theorem; and indeed its proof is analogous to Hyers's. Furthermore we can replace the vector space X by an abelian group G, without any change in the proof.

P. W. Cholewa independently proved in [17], together with some results about approximately convex functions, the same Theorem 14, but with a longer proof.

In [30] I. Fenyö improved the bound obtained by Skof and Cholewa: he showed that (11) holds with $\varepsilon = \delta/3 + \|f(0)\|/3$ (note that this bound depends not only on δ).

As for the Cauchy equation (A), Skof considered the local stability of the quadratic equation and, by using extension techniques, she was able to prove in [95] the following theorem.

THEOREM 15 (F. Skof). *Let $f\colon [0, a) \to Y$, $a > 0$, and Y be a Banach space and define*

$$K = \{(x, y) \in \mathbb{R}^2 \colon x \geq y \geq 0,\, x + y < a\}.$$

Assume that f satisfies (10) for $(x, y) \in K$ and, for

$$g(t) = f\left(\frac{a}{4} + t\right) - f\left(\left|\frac{a}{4} - t\right|\right), \qquad t \in \left[0, \frac{a}{2}\right],$$

suppose that

$$g(t_1 + t_2) = g(t_1) + g(t_2), \qquad for \ t_1, t_2 \in \left[0, \frac{a}{4}\right].$$

Then there exists a $\kappa: \mathbb{R} \to Y$ *satisfying* (Q), *such that*

$$\|f(x) - \kappa(x)\| \le \frac{81}{2}\,\delta, \qquad x \in [0, a).$$

Other results of this kind are in [96].

Now, as for the Cauchy equation, we consider the inequality

$$\|f(x + y) + f(x - y) - 2f(x) - 2f(y)\| \le \Lambda(x, y). \tag{12}$$

The following has been proved by St. Czerwik in [20].

THEOREM 16 (St. Czerwik). *Let* X *be a normed space and* Y *a Banach space and let* $f: X \to Y$ *be a function satisfying inequality* (12) *with either*
   (i) $\Lambda(x, y) = \eta + \theta(\|x\|^s + \|y\|^s)$, $s < 2$, $x, y \in X \backslash \{0\}$,
or
   (ii) $\Lambda(x, y) = \theta(\|x\|^s + \|y\|^s)$, $s > 2$, $x, y \in X$,
for some $\eta, \theta \ge 0$. *Then there exists exactly one solution* κ *of* (Q) *such that*

$$\|f(x) - \kappa(x)\| \le \frac{1}{3}(\eta + \|f(0)\|) + \frac{2\theta}{4 - 2^s}\|x\|^s, \qquad x \in X \backslash \{0\}$$

in case (i) *and*

$$\|f(x) - \kappa(x)\| \le \frac{2\theta}{2^s - 4}\|x\|^s, \qquad x \in X$$

in case (ii). *Moreover, if the function* $t \mapsto f(tx)$, $t \in \mathbb{R}$, *is continuous for each* $x \in X$, *then* κ *satisfies the equation*

$$\kappa(tx) = t^2\kappa(x), \qquad x \in X, \qquad t \in \mathbb{R}.$$

Theorem 16 contains as particular case a result by H. Drljević ([27]).

In [10] as a particular case of a stability theorem for a wider class of functional equations the following result has been obtained.

THEOREM 17 (C. Borelli and G. L. Forti). *Let $(G, +)$ be an abelian group, Y a Banach space and let $f: G \to Y$ be a function with $f(0) = 0$ and fulfilling (12). Assume that one of the series*

$$\sum_{i=1}^{+\infty} 2^{-2i}\Lambda(2^{i-1}x, 2^{i-1}x) \quad and \quad \sum_{i=1}^{+\infty} 2^{2(i-1)}\Lambda(2^{-i}x, 2^{-i}x)$$

converges for every x and call $\Gamma(x)$ its sum. If, for every x, y, as $i \to \infty$, $2^{-2i}\Lambda(2^{i-1}x, 2^{i-1}y) \to 0$ or $2^{2(i-1)}\Lambda(2^{-i}x, 2^{-i}y) \to 0$, respectively, then there exists a unique solution κ of (Q) such that

$$\|f(x) - \kappa(x)\| \le \Gamma(x), \qquad x \in X.$$

The condition $f(0) = 0$ cannot be omitted in general; nevertheless, when Λ has, for instance, the form

$$\Lambda(x, y) = \theta(\|x\|^s + \|y\|^s), \qquad s > 0,$$

then the above condition is automatically satisfied.

Theorem 17 contains as special cases part (ii) of Czerwik's and the result proved by J. M. Rassias in [77].

In [24], [25] and [27] H. Drljević proved the analogue of Theorem 13 for the quadratic equation.

THEOREM 18 (H. Drjlević). *Let $f: X \to \mathbb{C}$, where X is a complex Hilbert space with $\dim X \ge 3$. Let $T: X \to X$ be a continuous selfadjoint operator with $\dim T(X) \ne 1, 2$. If*

$$|f(x + y) + f(x - y) - 2f(x) - 2f(y)| \le \theta[|(Tx, x)|^{p/2} + |(Ty, y)|^{p/2}]$$

for some $\theta \ge 0$, $p \in [0, 2)$ and for all $x, y \in X$ such that $(Tx, y) = 0$, then there exists a $\kappa: X \to \mathbb{C}$ quadratic on the pairs (x, y) with $(Tx, y) = 0$ and such that

$$|f(x) - \kappa(x)| \le \theta_1 |(Tx, x)|^{p/2}, \qquad x \in X,$$

for some constant θ_1.

As for the Cauchy equation, when $\Lambda(x, y) = \theta(\|x\|^2 + \|y\|^2)$ the usual technique to get stability cannot be applied and indeed St. Czerwik in [20], by modifying Gajda's example for the Cauchy equation, proved the existence of a function

$f: \mathbb{R} \to \mathbb{R}$ satisfying (12) with the above Λ and such that for any quadratic κ the quotient

$$\frac{|f(x) - \kappa(x)|}{|x|^2}, \qquad x \neq 0,$$

is unbounded.

In [48] R. Ger by using a different control function got the following.

THEOREM 19 (R. Ger). *Let $(G, +)$ be an abelian group and $(Y, \|.\|)$ a real n-dimensional normed space. Let $f: G \to X$ be a function satisfying the inequality (12) with*

$$\Lambda(x, y) = 2F(x) + 2F(y) - F(x + y) - F(x - y),$$

where $F: G \to \mathbb{R}$ is a given non-negative function with

$$2F(x) + 2F(y) - F(x + y) - F(x - y) \geq 0, \qquad x, y \in G.$$

Then there exists a quadratic $\kappa: G \to Y$ such that

$$\|f(x) - \kappa(x)\| \leq nF(x), \qquad x \in G.$$

Moreover, if

$$\liminf_{k \to +\infty} \frac{F(kx)}{k^2} = 0$$

for all $x \in G$, then κ is uniquely determined.

In the second part of this section we consider the stability of the Fréchet and polynomial equations.

Let G be a semigroup and Y a vector space; for a function $g: G \to Y$ define, as usual, the first difference as

$$\Delta_y g(x) = g(x + y) - g(x), \qquad y \in G,$$

and, recursively,

$$\Delta^n_{y_1,\dots,y_n} g(x) = \Delta_{y_n} \{\Delta^{n-1}_{y_1,\dots,y_{n-1}} g(x)\}, \qquad y_1, \dots, y_n \in G.$$

We say that a function $\gamma: G \to Y$ satisfies the Fréchet equation of order n if

$$\Delta^n_{y_1,\ldots,y_n} \gamma(x) = 0, \qquad x, y_1, \ldots, y_n \in G. \tag{F_n}$$

A weaker form of equation (F_n) is obtained by assuming the increments $y_1, \ldots, y_n$ all equal to y; thus we have

$$\Delta^n_y \gamma(x) = 0, \qquad x, y \in G, \tag{P_n}$$

$(\Delta^n_y = \Delta^n_{y,\ldots,y})$.

A function γ satisfying (P_n) is called *polynomial of degree at most* $n - 1$. Clearly every solution of (F_n) is polynomial; moreover D. Ž. Djoković in [23] proved that if G is commutative then equations (F_n) and (P_n) are equivalent.

If $f: [0, 1] \to \mathbb{R}$ has derivatives up to the n-th order and $f^{(n)}$ is bounded, then there exists a polynomial p of degree at most $n - 1$ such that

$$|f(x) - p(x)| \leq \frac{1}{n!} \sup_y |f^{(n)}(y)|, \qquad 0 \leq x \leq 1;$$

the polynomial p may be obtained for instance from Taylor's formula. Moreover, it is possible to replace $\sup_y |f^{(n)}(y)|$ by the upper bound of the n-th divided difference. H. Burkill in the 1950s conjectured that a similar result holds if we use the n-th differences instead.

In 1957 H. Whitney in his paper *On functions with bounded n-th differences* ([118]) answered the question by proving the following.

THEOREM 20 (H. Whitney). *For each integer $n \geq 1$ there is a number K_n with the following property. Let I be any closed interval, then for any continuous function f in I there is a polynomial p of degree at most $n - 1$ such that*

$$|f(x) - p(x)| \leq K_n \sup_{y,h} |\Delta^n_h f(y)|, \qquad x \in I.$$

Whitney takes as p the Lagrange polynomial which is equal to f at the n points of division of I in $n - 1$ equal intervals. This polynomial does not give the best approximation in the sense of minimizing

$$\sup_{x \in I} |f(x) - p(x)|.$$

J. C. Burkill modified the definition of n-th difference and in [13] proved the analogue of Theorem 20 by using the best approximating polynomial.

One year later Whitney returned to the problem and in [119] got the same conclusion as in Theorem 20 but replacing the continuity of f by the boundedness of f on an interval $I' \subset I$. All these results are about *ordinary* polynomials.

A notable generalization has been achieved by D. H. Hyers in [54] by using the abstract notion of polynomial given by Fréchet in [38].

Let X be a vector space over $\mathbb{Q}$ and Y be a vector space and denote by S a convex cone in X with vertex at the origin. We say that $\gamma\colon S \to Y$ is a *polynomial of degree at most $n-1$* if it is a solution of equation (P_n) where $x, y \in S$. A function $\gamma_k\colon S \to Y$ is called *rational-homogeneous form of degree k on S*, if it is not identically zero on S and

$$\gamma_k(x) = \gamma^*(x, \ldots, x),$$

where $\gamma^*\colon S^k \to Y$ is a k-additive function, i.e., is additive with respect to each variable.

S. Mazur and W. Orlicz proved in [65] and [66] that γ is a polynomial of degree at most $n-1$ on S if and only if

$$\gamma(x) = \gamma_0(x) + \gamma_1(x) + \cdots + \gamma_{n-1}(x), \tag{13}$$

where each γ_k is a rational-homogeneous form of degree k on S.

The main result of Hyers's paper [54] reads as follows.

THEOREM 21 (D. H. Hyers). *Let S be a convex cone in a rational vector space X and let Y be a Banach space. Let $f\colon S \to Y$ be a function satisfying the inequality*

$$\left\| \Delta^n_{y_1,\ldots,y_n} f(x) \right\| \leq \delta, \qquad x, y_1, \ldots, y_n \in S, \tag{14}$$

for some $\delta \geq 0$. Then there exists a polynomial γ of degree at most $n-1$ such that

$$\| f(x) - \gamma(x) \| \leq \delta, \qquad x \in S.$$

Moreover γ is given by the formula (33) *where $\gamma_0(x) = f(0)$,*

$$\gamma_{n-1}(x) = \lim_{s \to +\infty} \frac{1}{(n-1)!s^{n-1}} \Delta^{n-1}_{sx} f(0)$$

$$\gamma_k(x) = \lim_{s \to +\infty} \frac{1}{k!s^k} \left\{ \Delta^k_{sx} f(0) - \sum_{j=k+1}^{n-1} \Delta^k_{sx} \gamma_j(0) \right\}, \qquad \text{for } 1 \leq k \leq n-2.$$

The polynomial γ is unique up to a constant. If X is a normed space and $S = X$, then the boundedness of f on a non-empty open subset of X implies the continuity of γ on X.

In view of the result of Djoković about the equivalence of (F_n) and (P_n) (note that this result appeared eight years later), Theorem 21 is a stability result for the Fréchet equation.

Whitney's results for unbounded intervals are contained in those of Hyers; this is not true anymore for the case of bounded intervals. Moreover, while Whitney uses the boundedness of the n-th difference with equal increments, Hyers uses independent increments; a stronger condition.

More than twenty years after the publication of Hyers's theorem, M. Albert and J. A. Baker in [2] generalized and gave a short proof of Hyers's theorem and proved a more general form of Whitney's results; they used the theorem proved by Djoković in [23].

Their results are summarized in the following two theorems.

THEOREM 22 (M. Albert and J. A. Baker). *Let G be a commutative semigroup and Y a Banach space. Let $f: G \to Y$ be a function satisfying inequality* (14) *for some $\delta \geq 0$. Then there exist symmetric, k-additive $\gamma_k^*: G^k \to Y$, $1 \leq k \leq n-1$, such that*

$$\left\| \Delta_y \left(f - \sum_{k=1}^{n-1} \gamma_k \right)(x) \right\| \leq \delta, \qquad x, y \in G, \tag{16}$$

where $\gamma_k(x) = \gamma_k^*(x, \ldots, x)$.

If G has a zero, then from (16) *we obtain* (15).

THEOREM 23 (M. Albert and J. A. Baker). *Let G be a commutative semigroup and Y a Banach space. Let $f: G \to Y$ be a function satisfying the inequality*

$$\| \Delta_y^n f(x) \| \leq \delta, \qquad x, y \in G,$$

for some $\delta \geq 0$.

Then inequality (16) *holds with $M_n \delta$ instead of δ, where M_n is a constant depending only on n.*

If G is divisible by $n!$, then we can take $M_n = 2$; if G is a group divisible by $n!$, then

$$M_n = \frac{2}{\sup\limits_{m} \dbinom{n}{m}}$$

If, moreover, G is a cone with non-empty interior in a normed space X and f is bounded on a non-empty open subset of G, then $\gamma_1, \ldots, \gamma_{n-1}$ *are continuous.*

Theorem 22 contains Hyers's theorem, Theorem 23 generalizes Whitney's results.

L. Székelyhidi in the already quoted paper [102] gave a new proof of Theorem 22 (for $Y = \mathbb{C}$) by using invariant means.

It is natural to try to extend the previous result to amenable semigroups. The problem is that Djoković's result only guarantees that the solutions of the Fréchet equation (F_n) are representable in the form (13) for commutative semigroups. Hence in the following stability theorem we can only say that the function γ which appears is a solution of (F_n).

THEOREM 24 (L. Székelyhidi [105]). *Let G be an amenable semigroup with identity and let* $f: G \to \mathbb{C}$ *be a function satisfying* (14) *for some* $\delta \geq 0$. *Then there exists a solution* $\gamma: G \to \mathbb{C}$ *of* (F_n) *for which* $f - \gamma$ *is bounded.*

It should be noted that in the same paper [105] it is proved that under the additional condition

$$\gamma(t + x + y) = \gamma(t + y + x),$$

for all $t, x, y \in G$, the solutions of (F_n) still have the form (13).

We mention here, without details, a result of J. Schwaiger about the stability of an equation, similar to a "finite difference Taylor expansion", characterizing the homogeneous polynomials on a commutative group ([89]).

Following the results of R. Ger stated in Theorem 10, P. W. Cholewa in [19] proved the following about the stability of the polynomial equation on restricted domain.

THEOREM 25 (P. W. Cholewa). *Let* $n \in \mathbb{N}$ *be fixed and let G be an abelian group in which the division by* $n!$ *is uniquely performable. Let* $\mathscr{F}_1$ *and* $\mathscr{F}_2$ *be conjugate p.l.i.* σ*-ideals in G and* $G \times G$ *respectively and assume that*

$$A \in \mathscr{F}_i \text{ implies } \frac{1}{k} A \in \mathscr{F}_i, \quad k = 1, 2, \ldots, n, \quad i = 1, 2.$$

Let $f: G \to Y$ *(Y Banach space) be a function satisfying the inequality*

$$\|\Delta_y^n f(x)\| \leq \delta, \quad (x, y) \in G^2 \backslash M,$$

for some $M \in \mathscr{F}_2$ *and* $\delta \geq 0$.

Then there exist a solution $\gamma: G \to Y$ *of* (P_n), *a constant* k_n *and a set* $E \in \mathscr{F}_1$ *such that*

$$\|f(x) - \gamma(x)\| \le k_n \delta, \qquad x \in G \backslash E.$$

Z. Gajda in [40] obtained some results of local stability for the equations (F_n) and (P_n), analogous to those proved by F. Skof in [93] and [95] for the Cauchy and quadratic equation.

After proving some results about the local stability of multiadditive functions (using extension techniques), Gajda got the following.

THEOREM 26 (Z. Gajda). *Let* δ, $a \in (0, +\infty)$, $n \in \mathbb{N}$ *and let* Y *be a Banach space. If a function* $f: [0, a) \to Y$ *satisfies the inequality* (14) *for all* $x, y_1, \ldots, y_n \in [0, a)$ *such that* $x + y_1 + \cdots + y_n < a$, *then there exists a solution* $\gamma: \mathbb{R} \to Y$ *of* (F_n) *with the property*

$$\|f(x) - \gamma(x)\| \le k_n \delta, \qquad x \in [0, a),$$

where k_n *is a constant depending only on* n. *If* $f: (-a, a) \to Y$ *satisfies the inequality* (14) *for all* $x, y_1, \ldots, y_n \in (-a, a)$ *such that* $|x| + |y_1| + \cdots + |y_n| < a$, *then there exists a solution* $\gamma: \mathbb{R} \to Y$ *of* (F_n) *with the property*

$$\|f(x) - \gamma(x)\| \le k_n' \delta, \qquad x \in (-a, a),$$

where k_n' *is a constant depending only on* n.

THEOREM 27 (Z. Gajda). *Let* $x_0 \in \mathbb{R}$, δ, $a \in (0, +\infty)$, $n \in \mathbb{N}$ *and let* Y *be a Banach space. If a function* $f: (x_0 - a, x_0 + a) \to Y$ *satisfies the inequality*

$$\|\Delta_y^n f(x)\| \le \delta \tag{17}$$

for all $x \in (x_0 - a, x_0 + a)$ *and* $y \in (-a, a)$ *such that* $x + ny \in (x_0 - a, x_0 + a)$, *then there exists a solution* $\gamma: \mathbb{R} \to Y$ *of* (P_n) *such that*

$$\|f(x) - \gamma(x)\| \le m_n \delta, \qquad x \in (x_0 - a, x_0 + a),$$

where m_n *is a constant depending only on* n. *If* $r > 0$ *and* $f: (-r\sqrt{1+n^2}, r\sqrt{1+n^2}) \to Y$ *satisfies* (17) *for all* $(x, y) \in \mathbb{R}^2$ *with* $x^2 + y^2 < r^2$, *then there exists a solution* $\gamma: \mathbb{R} \to Y$

of (P_n) *such that*

$$\|f(x) - \gamma(x)\| \le m'_n \delta, \qquad x \in (-r\sqrt{1+n^2}, r\sqrt{1+n^2}),$$

where m'_n is a constant depending only on n.

To conclude this section we mention a paper of L. Székelyhidi ([98]) concerning the stability of a linear functional equation on abelian groups, whose solutions are polynomials.

4. The multiplicative Cauchy equation. Superstability

In Section 2 we considered the Cauchy equation (A) and the additive notation on the right-hand side was intended to emphasize that the range of the function is the additive group of a vector space.

If the range has the richer algebraic structure of an algebra, we can consider the multiplicative equation

$$m(xy) = m(x)m(y) \qquad\qquad (M)$$

and we may ask what can be said about a function f for which the difference

$$f(xy) - f(x)f(y)$$

is bounded in some sense to be specified.

Thus we are led to the problem of studying the stability of equation (M). This investigation highlighted a new phenomenon which is now usually called **superstability**. Before giving a definition of superstability, we present here the first result obtained by D. G. Bourgin in 1949 in connection with approximate isometrices.

THEOREM 28 (D. G. Bourgin [11]). *Let S_1, S_2 be compact and let $C(S_1)$, $C(S_2)$ be the algebras of real continuous functions on S_1 and S_2 respectively, with the sup norm. Assume $f: C(S_1) \to C(S_2)$ satisfies the inequality*

$$\|f(xy) - f(x)f(y)\| \le \delta, \qquad x, y \in C(S_1), \qquad\qquad (18)$$

for some $\delta \ge 0$ and moreover suppose that for each $s \in S_1$ there is a set $I(s) \subset \mathbb{R}$ such that

$$\sup_{\lambda \in I(s)} f(\lambda e)(s) = +\infty$$

(e is the function identically 1 on S_1). Then f is multiplicative, i.e., is a solution of (M).

Thus under the hypotheses of Theorem 28, the inequality (18) implies the equality (M).

Years later the problem of studying the stability of (M) in its generality has been reproposed by E. Lukacs in connection with some questions in probability.

An answer came with the following theorem (as in Section 2 when the domain is a vector space we use additive notation on the variables).

THEOREM 29 (J. A. Baker, J. Lawrence and F. Zorzitto [6]). *Let X be a $\mathbb{Q}$-vector space and let $f: X \to \mathbb{R}$ be a function such that*

$$|f(x+y) - f(x)f(y)| \le \delta, \qquad x, y \in X, \tag{19}$$

for some $\delta \ge 0$.

Then either $|f(x)| \le \max(4, 4\delta)$, or f is a solution of equation (M), i.e.,

$$f(x+y) = f(x)f(y), \qquad x, y \in X.$$

We can now give the following definition.

DEFINITION 3. Consider the functional equation

$$E(\varphi) = 0$$

and assume we are in a framework where the notion of boundedness of φ and of $E(\varphi)$ makes sense. Moreover, suppose that $E(\varphi)$ is bounded for any bounded φ.

We say that the equation above is **superstable** if the boundedness of $E(\varphi)$ implies that either φ is bounded or $E(\varphi) = 0$, i.e., φ is a solution of the equation.

Thus Theorems 28 and 29 say that, in the two different situations considered, equation (M) is superstable.

Some more comments are necessary about Definition 3. Since the function identically zero is multiplicative, i.e., is a solution of (M), the superstability of (M) is a special case of stability: the function f satisfying (19) differs from a solution of (M) by a bounded function, the zero function or f itself.

This remark can be extended to all functional equations $E(\varphi) = 0$ which are superstable and have the zero-function as a solution.

We already met in Section 2 some *trivial* superstability theorems: the additive Cauchy equation (A) is superstable when the domain is a finite group or semigroup and the range a normed space. But in this case superstability is due to the fact that the only solution of (A) is the function identically zero. So in the following we are interested in non-trivial superstability.

It should be noted that perhaps the first superstability result has been proved by H. N. Shapiro in [90] for a problem in number theory coming from a conjecture of Ulam.

In the first part of this section we confine ourselves to further results about equation (M).

A year after the publication of Theorem 29, J. A. Baker extended and greatly simplified it.

THEOREM 30 (J. A. Baker [5]). *Let G be a semigroup and $f: G \to \mathbb{C}$ be a function satisfying the inequality*

$$|f(xy) - f(x)f(y)| \le \delta, \qquad x, y \in G, \tag{20}$$

for some $\delta \ge 0$.

Then either $|f(x)| \le (1 + \sqrt{1 + 4\delta})/2$ or f is multiplicative.

Baker pointed out that the crucial step in the proof is the fact that $|zw| = |z||w|$ for all complex z and w. Indeed Theorem 30 holds true for functions with values in normed algebras for which the norm is multiplicative, like, for instance, the quaternions or the Cayley numbers. Moreover he gave an example of an unbounded function $f: \mathbb{R} \to M_2(\mathbb{C})$ ($M_2(\mathbb{C})$ is the algebra of 2×2 complex matrices with the usual norm) for which (20) holds, but it is not true that $f(x + y) = f(x)f(y)$ for all $x, y \in \mathbb{R}$.

Note that two years later essentially the same result (with a group instead of a semi-group) was proved by A. L. Shtern in [91].

R. Ger in [47] got the same result as in Theorem 30 (up to the numerical value of the bounding constant) by requiring only the boundedness (not necessarily uniform) of the function $x \mapsto f(xy) - f(y)$ for each $y \in G$.

From Theorem 30 we infer that, while for the stability of the additive Cauchy equation (A) the crucial role is played by the properties of the domain of the functions involved, for the superstability of the multiplicative equation (M) the crucial role is played by the range; concerning the domain we simply require it to be a semigroup.

R. Ger in [49] noted that this phenomenon of superstability is caused by the fact that we "mix" the two operations in $\mathbb{C}$: on the right-hand side of (M) we have a product and we measure the distance between the two sides of the equation by using the difference. If we use, in some more natural sense, the quotient we again obtain the traditional stability.

Baker's example regarding matrix-valued functions opened the studies on the stability of (M) for functions with values in $M_n(\mathbb{C})$, the algebra of the $n \times n$ complex matrices (see [1] and [120]).

J. Lawrence in [64] proved the following results on superstability and stability.

THEOREM 31 (J. Lawrence). *Let G be a semigroup, $\mathcal{T}$ an associative normed algebra with unit. Let $f: G \to \mathcal{T}$ satisfying the inequality (18) for all $x, y \in G$ and for some $\delta \geq 0$.*

If the subalgebra $\mathcal{A} \subset \mathcal{T}$ generated by the range of f is simple, then either f is bounded or f is multiplicative.

In particular if $\mathcal{T} = \mathcal{A} = M_n(\mathbb{C})$, then f is bounded or multiplicative.

THEOREM 32 (J. Lawrence). *if G is an abelian group and $f: G \to M_2(\mathbb{C})$ satisfies* (18), *then there exists a solution m of (M) such that $f - m$ is bounded.*

Theorem 32 was later extended to $M_n(\mathbb{C})$ by D. Dicks (see [22]).

Now we go back to Theorems 29 and 30: L. Székelyhidi, while proving in [100] a generalization of these theorems, noted the algebraic character of them. This fact led him to the following abstract superstability theorem.

THEOREM 33 (L. Székelyhidi [107]). *Let H be a module over the division ring R, let $\Theta \subset$ End H be a semigroup, $H_0 \subset H$ a Θ-invariant submodule and $f: \Theta \to R$ a function.*

If $h \in H$ is such that

$$\xi(h) - f(\xi)h \in H_0 \qquad \text{for all } \xi \in \Theta, \tag{21}$$

then either $h \in H_0$ or f is an antihomomorphism of Θ.

Theorem 30 follows from Theorem 33 by taking a semigroup G, $R = \mathbb{C}$, H as the module of all complex functions on G, $\Theta = G$, H_0 as the set of bounded complex functions and, if $x \in G$, the corresponding endomorphism is the left translation. Thus, taking $\xi - x$ and $h - f$ in (21), we obtain that the boundedness of $f(xy) - f(x)f(y)$ implies that f is bounded or multiplicative.

In Theorems 30 and 31 the superstability of (M) has been proved without any condition on the function f and by assuming the greatest generality for the domain. As already noted, the crucial role is played by the properties of the range.

If we require some additional condition on f and specialize the domain (as for instance in Theorem 28) we can again get superstability also for more general spaces of the values. One of the most recent results in this sense is the following.

THEOREM 34 (P. Šemrl [87]). *Let X, Y be Banach spaces and let $\mathcal{A}$ and $\mathcal{B}$ be standard operator algebras on X and Y respectively (an algebra of bounded operators* *is called standard if it contains the bounded finite-rank operators). Let $f: \mathscr{A} \to \mathscr{B}$ be a bijective mapping satisfying (18) for all $x, y \in \mathscr{A}$ and for some $\delta \geq 0$. Then f is multiplicative.*

Till now all functions had their range in a field or in an algebra, so among the solutions of (M) there is the identically zero function.

Now consider a group G, a complex Hilbert space $\mathfrak{H}$ and let $GL(\mathfrak{H})$ be the group of all bounded invertible operators on $\mathfrak{H}$, endowed with the norm topology. A representation τ of G into $\mathfrak{H}$ is a homomorphism $\tau: G \to GL(\mathfrak{H})$, i.e., a solution of (M).

It is natural to pose the stability problem for such representations and it has been treated by P. de la Harpe and M. Karoubi; in [21] they proved the following.

THEOREM 35 (P. de la Harpe and M. Karoubi). *Let G be a compact topological group and let k, ε be two real numbers with $k > 1$ and $\varepsilon \geq 0$.*

Then there exists $\delta > 0$ with the following properties.

For all functions $f: G \to GL(\mathfrak{H})$, satisfying (20) and such that $\|f(x)\| \leq k$ and $\|f(x)^{-1}\| \leq k$ for every $x \in G$, there exists a continuous representation $\tau: G \to GL(\mathfrak{H})$ with

$$\|f(x) - \tau(x)\| \leq \varepsilon, \qquad x \in G.$$

For the same problem D. Kazhdan proved in [62] the following theorem.

THEOREM 36 (D. Kazhdan). *Let G be a topological amenable group and let U be the group of unitary transformations of a Hilbert space $\mathfrak{H}$. Assume that $f: G \to U$ is a function satisfying (20) with $\delta < 1/100$. Then there exists a representation $\tau: G \to U$ such that*

$$\|f(x) - \tau(x)\| \leq \delta, \qquad x \in G.$$

Some generalizations of these results are proved in [92] by A. I. Shtern.

In the second part of this section we present superstability results concerning other functional equations.

J. A. Baker in the previously quoted paper [5] considered the *cosine equation*

$$c(xy) + c(xy^{-1}) = 2c(x)c(y), \tag{CS}$$

where x and y are in an abelian group G and the function c is complex valued.

He proved the following.

THEOREM 37 (J. A. Baker). *Let G be an abelian group and let $f: G \to \mathbb{C}$ be a function such that*

$$|f(xy) + f(xy^{-1}) - 2f(x)f(y)| \le \delta, \qquad x, y \in G, \tag{22}$$

for some $\delta \ge 0$. Then either $|f(x)| \le (1 + \sqrt{1 + 2\delta})/2$ for all $x \in G$ or f is a solution of (CS).

Thus equation (CS) is superstable for complex functions. To prove Theorem 37, Baker uses the well known fact that c is a solution of equation (CS) if and only if there exists a multiplicative function m such that $c(x) = \{m(x) + m(-x)\}/2$. Then the result is a consequence of Theorem 30.

It should be noted that this was the first time that for a stability result we needed to know the form of the solutions of the equation.

Theorem 37 has been extended by L. Székelyhidi:

THEOREM 38 (L. Székelyhidi [99]). *Let G be a group, F a field and V a vector space of functions from G into F, invariant by right translation.*

Let $f, g: G \to F$ be functions such that

$$f(xyz) = f(xzy), \qquad x, y, z \in G,$$

and for every $y \in G$ the function

$$x \mapsto f(xy) + f(xy^{-1}) - 2f(x)g(y)$$

belongs to V.

Then either $f \in V$ or g is a solution of (CS).

Taking V as the space of bounded functions and $f = g$, we get Theorem 37. Other generalizations of equation (CS) were studied in [101].

The *sine equation*

$$s(xy)s(xy^{-1}) = s(x)^2 - s(y)^2 \tag{SN}$$

has been studied by P. W. Cholewa who, in [16], showed that it is superstable for complex .functions, provided x, y belong to a uniquely 2-divisible commutative group.

Furthermore P. W. Cholewa considered in [18] a general equation of the form

$$\phi(F(x, y)) = H(\phi(x), \phi(y))$$

and, under suitable hypotheses on the functions and the spaces involved, was able to prove some superstability theorems generalizing those concerning the multiplicative equation.

5. Approximately multiplicative linear maps in Banach algebras

About twenty years ago some papers began to appear dealing with *perturbations* of Banach algebras. Essentially three kinds of perturbations are considered: perturbations of the multiplicative structure, of isometrices and of homomorphisms (see the book of K. Jarosz [58]).

In the present paper we are interested in perturbations of homomorphisms and we start with the following definition.

DEFINITION 4. Let $\mathfrak{A}$ and $\mathfrak{B}$ be two Banach algebras. Given $\delta \geq 0$, a linear map T from $\mathfrak{A}$ into $\mathfrak{B}$ is called δ-*multiplicative* (or δ-homomorphism) if

$$\|T(xy) - T(x)T(y)\| \leq \delta \|x\| \|y\|, \qquad x, y \in \mathfrak{A}. \tag{23}$$

We are interested in whether there exists a homomorphism *near* T, i.e., we are studying a generalized stability problem for the multiplicative Cauchy equation restricted to linear maps.

The considerations of the above problem will be divided in two different cases: the case of functionals, i.e., $\mathfrak{B} = \mathbb{C}$, and the general case.

First we consider the approximately multiplicative functionals, i.e., $\mathfrak{B} = \mathbb{C}$.

A result of K. Jarosz ([58]) says that if $T: \mathfrak{A} \to \mathbb{C}$ satisfies (23), then $\|T\| \leq 1 + \delta$ and so it is a continuous linear functional. Since in the case of functionals it is natural to restrict the considerations to commutative Banach algebras, we assume $\mathfrak{A}$ commutative.

Denote by $\hat{\mathfrak{A}}$ the set of characters, that is the non-zero multiplicative linear functionals on $\mathfrak{A}$ and by $\mathfrak{A}^*$ the dual of $\mathfrak{A}$. For each $T \in \mathfrak{A}^*$ we put

$$d(T) = \inf\{\|T - S\| : S \in \hat{\mathfrak{A}} \cup \{0\}\}.$$

Following B. E. Johnson ([60]) we give the

DEFINITION 5. The Banach algebra $\mathfrak{A}$ is *an algebra in which approximately multiplicative functionals are near multiplicative functionals*, or $\mathfrak{A}$ is AMNM for short, if for each $\varepsilon > 0$ there is a $\delta > 0$ such that $d(T) < \varepsilon$ whenever T is a δ-multiplicative linear functional, i.e., there exists $S \in \hat{\mathfrak{A}} \cup \{0\}$ such that $\|T(x) - S(x)\| < \varepsilon \|x\|, x \in \mathfrak{A}$.

In the following theorem we summarize some of the results proved in [60] by B. E. Johnson.

THEOREM 39 (B. E. Johnson). *The following commutative Banach algebras are AMNM:*
  (i) *any finite dimensional Banach algebra;*
  (ii) $C_0(X)$, *where X is a locally compact Hausdorff space;*
  (iii) l^p, $1 \leq p \leq \infty$;
  (iv) $L^1(G)$, *where G is a locally compact abelian group;*
  (v) *the algebra $l^1(\mathbb{Z}^+)$ of power series $\sum_{n=0}^{+\infty} a_n z^n$ with $\sum |a_n| < \infty$;*
  (vi) *the convolution algebra $L^1(0, +\infty)$;*
  (vii) *the disc algebra, that is, the algebra of the continuous functions on the closed unit disc in $\mathbb{C}$, analytic in the interior of the disc.*
      (l^p *is the usual algebra of sequences with multiplication* $\{a_n\}\{b_n\} = \{a_n b_n\}$.)

At the end of his interesting paper, Johnson gives an example of a non-AMNM algebra.

Now we return to the general case of maps between two Banach algebras $\mathfrak{A}$ and $\mathfrak{B}$. The situation is much more complicated. First of all, the continuity result of Jarosz in this more general case is not true. This implies that we have to assume the boundedness of the linear maps involved. Thus we denote by $L(\mathfrak{A}, \mathfrak{B})$ the space of bounded linear maps from $\mathfrak{A}$ into $\mathfrak{B}$ and by $M(\mathfrak{A}, \mathfrak{B})$ the subset of $L(\mathfrak{A}, \mathfrak{B})$ consisting of multiplicative maps. For $T \in L(\mathfrak{A}, \mathfrak{B})$ we put

$$d(T) = \inf\{\|T - S\| : S \in M(\mathfrak{A}, \mathfrak{B})\}.$$

DEFINITION 6. We say that $(\mathfrak{A}, \mathfrak{B})$ is an AMNM pair if for each positive ε and K there is $\delta > 0$ such that, if $T \in L(\mathfrak{A}, \mathfrak{B})$ with $\|T\| < K$ and satisfies (23), then $d(T) < \varepsilon$, i.e., there exists $S \in M(\mathfrak{A}, \mathfrak{B})$ such that $\|T(x) - S(x)\| < \varepsilon \|x\|$, $x \in \mathfrak{A}$.

B. E. Johnson studied the problem of AMNM pairs in [61]. The first result is that $(\mathfrak{A}, \mathfrak{B})$ is AMNM for all finite-dimensional $\mathfrak{A}$ and $\mathfrak{B}$.

In order to state the main result about AMNM pairs we need the definitions of Banach bimodule and of amenable algebra.

Let $\mathfrak{A}$ be a complex Banach algebra. A complex Banach space $\mathfrak{X}$ is a Banach $\mathfrak{A}$-bimodule if it is a two-sided $\mathfrak{A}$-module and there is a positive H such that

$$\|ax\| \leq H\|a\|\|x\| \quad \text{and} \quad \|xa\| \leq H\|x\|\|a\| \qquad \text{for all } a \in \mathfrak{A}, \quad x \in \mathfrak{X}.$$

When $\mathfrak{X}$ is a Banach $\mathfrak{A}$-bimodule, we define

$$\mathscr{L}_n(\mathfrak{A}, \mathfrak{X}) = \mathfrak{A} \hat{\otimes} \cdots \hat{\otimes} \mathfrak{A} \hat{\otimes} \mathfrak{X},$$

where there are n copies of $\mathfrak{A}$; $\mathscr{L}_n(\mathfrak{A}, \mathfrak{X})$ is the completion of the tensorial product $\mathfrak{A} \otimes \cdots \otimes \mathfrak{A} \otimes \mathfrak{X}$ endowed with the tensorial product norm (see for instance [52]). The dual space $\mathfrak{X}^*$ of $\mathfrak{X}$ becomes a Banach $\mathfrak{A}$-bimodule if we define

$$\langle ya, x \rangle = \langle y, ax \rangle, \qquad \langle ay, x \rangle = \langle y, xa \rangle, \qquad a \in \mathfrak{A}, \quad x \in \mathfrak{X}, \quad y \in \mathfrak{X}^*.$$

The dual space of $\mathscr{L}_n(\mathfrak{A}, \mathfrak{X})$ is the space of bounded multilinear functionals, n-linear in $\mathfrak{A}$ and linear in $\mathfrak{X}$. Given such a multilinear functional we have, for $a_1, \ldots, a_n \in \mathfrak{A}$, an element of $\mathfrak{X}^*$. So $\mathscr{L}_n(\mathfrak{A}, \mathfrak{X})^*$ can be identified with the space $\mathscr{L}^n(\mathfrak{A}, \mathfrak{X}^*)$ of the n-linear transformations of $\mathfrak{A}$ into $\mathfrak{X}^*$.

Now we form the complex

$$0 \xrightarrow{\delta^0} \mathfrak{X}^* \xrightarrow{\delta^1} \mathscr{L}^1(\mathfrak{A}, \mathfrak{X}^*) \xrightarrow{\delta^2} \mathscr{L}^2(\mathfrak{A}, \mathfrak{X}^*) \xrightarrow{\delta^3} \cdots,$$

where, for $T \in \mathscr{L}^{n-1}(\mathfrak{A}, \mathfrak{X}^*)$,

$$(\delta^n T)(a_1, \ldots, a_n) = a_1 T(a_2, \ldots, a_n) + \sum_{i=1}^{n-1} T(a_1, \ldots, a_i a_{i+1}, \ldots, a_n)$$

$$+ (-1)^n T(a_1, \ldots, a_{n-1}) a_n.$$

We have $\delta^{n+1} \delta^n = 0$ and we define

$$\mathscr{H}^n(\mathfrak{A}, \mathfrak{X}^*) = \frac{\mathrm{Ker}\, \delta^{n+1}}{\mathrm{Im}\, \delta^n}.$$

DEFINITION 7. An *amenable algebra* is a Banach algebra $\mathfrak{A}$ with $\mathscr{H}^1(\mathfrak{A}, \mathfrak{X}^*) = 0$ for all Banach $\mathfrak{A}$-bimodules $\mathfrak{X}$.

The relation with the usual amenability of groups is the following: a group G is amenable if and only if $L^1(G)$ is.

We can now state the main result about AMNM pairs of Banach algebras.

THEOREM 40 (B. E. Johnson [61]). *Let $\mathfrak{A}$ be an amenable algebra and suppose that $\mathfrak{B}$ is a Banach algebra such that there exists a Banach $\mathfrak{B}$-bimodule $\mathfrak{N}$ so that $\mathfrak{B}$ is isomorphic as $\mathfrak{B}$-bimodule with $\mathfrak{N}^*$. Then $(\mathfrak{A}, \mathfrak{B})$ is AMNM.*

From Theorem 40 we deduce the following.

THEOREM 41 (B. E. Johnson [61]). *In the following cases the pair* $(\mathfrak{A}, \mathfrak{B})$ *is AMNM*:

(*i*) $\mathfrak{A}$ *is a finite-dimensional semisimple algebra and* $\mathfrak{B}$ *is any Banach algebra*;

(*ii*) $\mathfrak{A} = l^1(\mathbb{Z})$ *and* $\mathfrak{B}$ *is commutative*;

(*iii*) $\mathfrak{A}$ *is the group algebra of a compact abelian group and* $\mathfrak{B}$ *is commutative*;

(*iv*) $\mathfrak{A} = \mathfrak{B} = L^1(\mathbb{R})$.

There is a (*non-commutative*) *Banach algebra* $\mathfrak{B}$ *such that* $(c_0, \mathfrak{B})$ *is not an AMNM pair.*

An interesting special situation is presented by the case $\mathfrak{B} = C(X)$, where X is a compact Hausdorff space. In this case Jarosz's theorem concerning the boundedness of approximately multiplicative functionals still holds; so in Definition 6 the condition $\|T\| < K$ can be omitted.

In the simplest case, i.e., X is a one-point space, $C(X) = \mathbb{C}$, so we can reformulate Definition 5 by saying that a commutative Banach algebra $\mathfrak{A}$ is an AMNM algebra if $(\mathfrak{A}, \mathbb{C})$ is an AMNM pair.

From Theorem 40 we get that, if a commutative Banach algebra $\mathfrak{A}$ is amenable, then it is AMNM.

For further results about AMNM pairs see [61].

6. Other equations and systems

In Section 4 we reported some superstability results concerning the cosine and sine functional equations (*CS*) and (*SN*). Obviously these equations are not the only ones satisfied by the functions cosine and sine. Thus a natural question arises whether the superstability phenomenon described for (*CS*) and (*SN*) carries over to other equations or systems satisfied by the trigonometric functions.

L. Székelyhidi in [109] considered the functional equations describing the additive theorem for sine and cosine, namely

$$s(xy) = s(x)c(y) + c(x)s(y) \tag{24}$$

$$c(xy) = c(x)c(y) - s(x)s(y), \tag{25}$$

where c, s are complex functions defined on a group G.

He proved that these two equations are stable, but not superstable. More exactly, if $f, g: G \to \mathbb{C}$ are given functions, G is an amenable group, and the function

$$(x, y) \mapsto f(xy) - f(x)g(y) - g(x)f(y)$$

is bounded, then there exists a solution (s, c) of (24) such that $f - s$ and $g - c$ are bounded.

An analogous result holds for equation (25).

R. Ger in [47], instead of considering equations (24) and (25) separately, studied the stability behaviour of the system

$$\begin{cases} s(xy) = s(x)c(y) + c(x)s(y) \\ c(xy) = c(x)c(y) - s(x)s(y) \end{cases} \tag{26}$$

Just as the single equations, also system (26) is not superstable, but it is stable.

THEOREM 42 (R. Ger). *Let G be a semigroup and let $h, k: G \times G \to [0, +\infty)$ be such that the sections $h(., x)$ and $k(x, .)$ are bounded for each $x \in G$. Assume $f, g: G \to \mathbb{C}$ satisfy the system of inequalities*

$$\begin{cases} |f(xy) - f(x)g(y) - g(x)f(y)| \leq h(x, y) \\ |g(xy) - g(x)g(y) + f(x)f(y)| \leq k(x, y) \end{cases}$$

for all $x, y \in G$. Then there exists a solution (s, c) of (26) such that $f - s$ and $g - c$ are bounded.

If, instead of complex valued functions, we assume $f, g: G \to \mathscr{A}$, where $\mathscr{A}$ is a normed algebra and G has a unit, we get the following:

THEOREM 43 (R. Ger). *Assume $f, g: G \to \mathscr{A}$ satisfy the system of inequalities*

$$\begin{cases} \|f(xy) - f(x)g(y) - g(x)f(y)\| \leq h(x, y) \\ \|g(xy) - g(x)g(y) + f(x)f(y)\| \leq k(x, y) \end{cases}$$

with h, k as in Theorem 42. If the subalgebra generated by $\{(f(x), g(x)): x \in G\}$ in the complexification $c(\mathscr{A})$ of $\mathscr{A}$ is simple, then either f and g are bounded or (f, g) is a solution of (26). Thus in this case system (26) is superstable.

As a consequence we have superstability if $\mathscr{A} = \mathbb{R}$.

Theorems 42 and 43 use in their proof a generalization of J. Lawrence's result contained in [64].

W. Förg-Rob and J. Schwaiger investigated the stability of the subtraction theorem for trigomometric and hyperbolic functions, i.e., the systems

$$\begin{cases} s(xy^{-1}) = s(x)c(y) - c(x)s(y) \\ c(xy^{-1}) = c(x)c(y) + s(x)s(y) \end{cases} \tag{27}$$

and

$$\begin{cases} S(xy^{-1}) = S(x)C(y) - C(x)S(y) \\ C(xy^{-1}) = C(x)C(y) - S(x)S(y), \end{cases} \tag{28}$$

where $s, c, S, C: \mathbb{C}$ and G is an abelian group.

They proved in [31] the superstability of both systems.

THEOREM 44 (W. Förg-Rob and J. Schwaiger). *Let $f, g: G \to \mathbb{C}$ satisfy*

$$\begin{cases} |f(xy^{-1}) - f(x)g(y) + g(x)f(y)| \le b(x) \\ |g(xy^{-1}) - g(x)g(y) - f(x)f(y)| \le b(x) \end{cases}$$

for some $b: G \to [0, +\infty)$. Then either f and g are bounded or the pair (f, g) is a solution of (27). *The same holds for system* (28).

The different behaviour of systems (26) and (28) led to a deeper investigation of these systems and in [32] W. Förg-Rob and J. Schwaiger studied a more general one and obtained results of stability and superstability containing those previously stated. Since the explanation of these results is quite long, the reader is referred to the original paper.

One of the most important functional equations is the *associativity equation*

$$F(x, F(y, z)) = F(F(x, y), z);$$

we assume here $F: \mathbb{R}^+ \times \mathbb{R}^+ \to \mathbb{R}^+$ and $F(x, 0) = x$, $x \in \mathbb{R}^+$; if we also want to have the homogeneity of the operation F, we are led to consider the equation

$$F(kx, kF(y, z)) = F(kF(x, y), kz), \qquad x, y, k \in \mathbb{R}^+. \tag{29}$$

What can be said about a binary operation F on $\mathbb{R}^+$ for which the two members of equation (29) are close? C. Alsina in [3] proved that (29) is superstable, i.e., if F

satisfies the inequality

$$|F(kx, kF(y, z)) - F(kF(x, y), kz))| \le \delta, \qquad x, y, k \in \mathbb{R}^+,$$

for some $\delta \ge 0$, then F satisfies (29).

Moreover he showed that, if we fix $k = 1$, i.e., we consider the associativity equation, superstability fails to hold. Still open is the problem of the stability of this equation.

Till now we considered the additive Cauchy equation (A) and the multiplicative one (M) separately. If we wish to study the homomorphisms of an algebra $\mathscr{A}$ into an algebra $\mathscr{B}$, we must consider the system of the two equations. What happens if the two equations are only approximately satisfied? An answer has been given by D. G. Bourgin who, in [11], proved the following.

THEOREM 45 (D. G. Bourgin). *Let $\mathscr{A}$ and $\mathscr{B}$ be normed algebras, $\mathscr{B}$ complete. Assume f from $\mathscr{A}$ onto $\mathscr{B}$ satisfies the system of inequalities*

$$\begin{cases} \|f(x + y) - f(x) - f(y)\| \le \delta_1 \\ \|f(xy) - f(x)f(y)\| \le \delta_2 \end{cases}$$

for some $\delta_1, \delta_2 \ge 0$. Then f is an algebraic isomorphism of $\mathscr{A}$ onto $\mathscr{B}$.

If $\mathscr{A} = C(S_1)$, $\mathscr{B} = C(S_2)$, S_1, S_2 compact spaces, then f is a homomorphism.

The analogous problem for the complex field, but using the quotient to measure the "distance" between the two members of the multiplicative equations, has been studied by J. Tabor, who proved that in this case we have neither superstability nor stability (see [115]).

Another important functional equation is that defining the *multiplicative derivations* in algebras:

$$d(xy) = xd(y) + d(x)y, \qquad (D)$$

where $d: \mathscr{A} \to \mathscr{B}$, $\mathscr{A} \subset \mathscr{B}$, $\mathscr{A}$ and $\mathscr{B}$ algebras.

P. Šemrl in [88] proved the following superstability result for (D).

THEOREM 46 (P. Šemrl). *Let X be a Banach space and let $\mathscr{A}$ be a standard operator algebra on X. Assume that $\psi: \mathbb{R}^+ \to \mathbb{R}^+$ is a function such that*

$$\lim_{t \to +\infty} \frac{\psi(t)}{t} = 0.$$

Suppose that $f: \mathscr{A} \to B(X)$ $(B(X)$ is the algebra of all bounded operators on $X)$ is a mapping satisfying the inequality

$$\|f(xy) - xf(y) - f(x)y\| \leq \psi(\|x\| \|y\|),$$

for all $x, y \in \mathscr{A}$. Then f is a solution of (D).

Among the functional equations which have on $\mathbb{R}$ the same solutions (up to a constant) as the additive Cauchy equation (A) is the so-called Hosszú's equation

$$h(x + y - xy) + h(xy) = h(x) + h(y). \tag{H}$$

Because of the stability of (A), it is natural to ask what happens for (H). This problem has been attacked by C. Borelli who in [9] proved the following.

THEOREM 47 (C. Borelli). *Let $f: \mathbb{R} \to \mathbb{R}$ be a function such that the map*

$$(x, y) \mapsto f(x + y - xy) + f(xy) - f(x) - f(y)$$

is bounded and let f_1 be the even part of f. Then there exists a solution h of (H) such that $f - h$ is bounded if and only if the function

$$(x, y) \mapsto f_1(x + y - xy) + f_1(xy) - f_1(x) - f_1(y)$$

is bounded.

To conclude this section we mention a result of M. Bean and J. A. Baker ([8]) concerning the stability of the functional analogue of the wave equation and those presented in a group of papers by L. Székelyhidi ([106], [108], [110]) concerning stability and/or superstability of some functional equations arising in problems of mathematical economics and dimensional analysis and for some kinds of functional equations in several variables like, for instance, the cocycle equation. All these results are based on those involving the additive and multiplicative Cauchy equation.

7. Final remarks and open problems

Hyers's Theorems 1 and 21 have a common background in the notion of inner algebraic inverse (see, for instance, [69]). This fact has been highlighted by I. Fenyö in [29]. He proved that Hyers's results can be deduced from the following theorem.

THEOREM 48 (I. Fenyö). *Let S be a (non-empty) set and Y a normed space. Denote by $\mathcal{G}_1$ and $\mathcal{G}_2$ the vector spaces of the functions from S into Y and from $S \times S$ into Y respectively. Let $A: \mathcal{G}_1 \to \mathcal{G}_2$ be a linear operator such that the only bounded function in its kernel is the zero-function and assume that A has an inner algebraic inverse M with $\|M\| \le k$. Then, for each $f \in \mathcal{G}_1$ with $\|(Af)(x, y)\| \le \delta$, $x, y \in S$, there exists a unique g in $\ker(A)$ such that*

$$\|f(x) - g(x)\| \le k\delta, \qquad x \in S.$$

This theorem suggests the possibility of looking at other stability results from this general and abstract point of view.

Theorem 48 and Hyers's results are the main tools for solving some kinds of alternative and inhomogeneous Cauchy functional equations (see [34] and [35]). This fact explains the interest in stability of many researchers working in the field of functional equations. It is quite surprising that functional equationists and people working on perturbations in Banach algebras and representations, developed their researches completely independently and unaware of each other.

The notion of stability of the additive Cauchy equation given in Definition 1 led to investigations of the semigroups and groups for which equation (A) is stable. The results presented in Section 2 (Theorems 2, 3 and 4) describe some classes of such groups and semigroups. A complete characterization of the groups and semigroups where (A) is stable seems far off.

Moreover, we remind that the original Ulam problem concerns metric groups. As already noted in Section 2, the stability problem for equation (A) in this more general and natural setting seems only to have been touched on.

Investigations in this direction could clarify the concept of stability and provide a tool for attacking some alternative functional equations in a more natural ambience.

If we consider the stability problem when the Cauchy difference is controlled by a non-constant function $\Phi(x, y)$ (see equation (5)), it would be interesting to develop completely new methods which can enlarge the class of the possible functions Φ considered in [10] and [33].

Moreover, in this case there are no results about the stability on restricted domains: this field seems almost unexplored.

Analogous questions arise for the quadratic and the polynomial equation.

Concerning the multiplicative Cauchy equation, in addition to the need of a better understanding of the superstability phenomenon, it would be interesting to study the inequality

$$\|f(xy) - f(x)f(y)\| \leq \Phi(x, y)$$

in appropriate settings. The investigations of B. E. Johnson concern the case in which f is linear and $\Phi(x, y) = \theta \|x\| \|y\|$. They suggest that investigations of the above inequality will lead deeply into functional analysis.

Johnson's papers give rise to the following question: instead of studying approximately multiplicative linear maps, consider approximately additive (or linear) and approximately multiplicative maps, i.e., the system

$$\begin{cases} \|f(x + y) - f(x) - f(y)\| \leq \delta_1(\|x\| + \|y\|) \\ \|f(xy) - f(x)f(y)\| \leq \delta_2\|x\| \|y\|. \end{cases}$$

We already know that the first inequality alone doesn't yield stability, but the behaviour of the system seems completely unknown.

The results presented in the previous sections can be divided into two classes: those whose proofs require knowledge of the form of the solution of the functional equation we are dealing with and the others. The first situation occurs, for instance, in Theorems 37, 42, 43 and 44. In the other cases, starting from a function f satisfying a certain inequality, we obtain, through an explicit procedure, a function g, then we prove in a direct way that g satisfies the functional equation under consideration. It is interesting to investigate whether these different features hide some deeper property.

It is much more important to interpret the previous remark from another point of view. All stability results whose proofs don't use knowledge of the solutions of the functional equation involved, may be read as existence theorems. This fact is particularly interesting when we deal with general classes of functional equations like, for instance, in [10], [18] and [33]. In these cases from a solution of an inequality we can deduce the existence and construct a solution of the equation. This method resembles the proof of the existence of a solution of the initial-point problem for a first-order differential equation, via the construction of approximate solutions.

Thus it is worthwhile to develop a method for constructing approximate solutions of a functional equation.

At the end of Section 1 we quoted some papers on various definitions of stability and on the use of different forms of a functional equation, and we promised to return to this question. In the original Ulam problem it is required that for every $\varepsilon > 0$ there is a $\delta > 0$ such that (1) implies (2). We noted that our Definition 1 of stability for the additive Cauchy equation is different; nevertheless the fact of using functions with values in a Banach space gives an *a posteriori* equivalence.

In more general situations this is no longer true. In this regard we will present a modification of an example contained in [7] and in [68]. This example concerns functions defined on a semigroup and with values in the same semigroup and shows simultaneously that the notion of stability coming from Ulam's problem and that most commonly used, i.e., boundedness of the distance between $f(xy)$ and $f(x) + f(y)$ implies boundedness of the distance between f and φ for some additive φ, are not equivalent, and, moreover, that it is not independent of the form we use for writing the functional equation.

Consider the multiplicative semigroup $\mathbb{R}^+$ of the non-negative real numbers with the metric $\varrho(x, y) = |e^{-x} - e^{-y}|$ and consider the Cauchy equation

$$\varphi(xy) = \varphi(x) + \varphi(y), \tag{30}$$

where $\varphi: \mathbb{R}^+ \to \mathbb{R}^+$. The zero function is the only solution of this equation.

Obviously, for any $f: \mathbb{R}^+ \to \mathbb{R}^+$ we have $\varrho(f(x) + f(y), f(xy)) < 1$ for all $x, y \in \mathbb{R}^+$ and $\varrho(f(x), 0) < 1$ for all $x \in \mathbb{R}^+$. Now we take $\varepsilon \in (0, 1)$; for any $\delta > 0$ choose $n \in \mathbb{N}$ such that

$$\frac{1}{n}\left(1 - \frac{1}{n}\right) \le \delta \quad \text{and} \quad 1 - \frac{1}{n} > \varepsilon$$

and define

$$f(x) = \begin{cases} 0, & x \ne 0 \\ \log n, & x = 0. \end{cases}$$

Thus $\varrho(f(x) + f(y), f(xy)) \le \delta$ for all $x, y \in \mathbb{R}$ and $\varrho(f(0), 0) > \varepsilon$; so the answer to Ulam's problem is negative. But if, instead of measuring the distance between $f(x) + f(y)$ and $f(xy)$, we measure that from $f(x) + f(y) - f(xy)$ to 0, from

$$\varrho(f(x) + f(y) - f(xy), 0) \le \delta, \qquad x, y \in \mathbb{R},$$

by putting $y = 0$ we get

$$\varrho(f(x), 0) \le \delta, \qquad x \in \mathbb{R},$$

i.e., a positive answer to Ulam's problem with $\delta = \varepsilon$.

This example shows that, when working in metric groups, it is essential to clarify what kind of stability we are looking for and what we understand by an *approximate* solution of a functional equation.

Fortunately, as is proved in [7], the case of the additive Cauchy equation with values in a normed space is particularly good: all these problems disappear.

Acknowledgment

The author is grateful to the Scientific Committee of the 30th International Symposium in Functional Equations, which invited him to present a survey talk on stability of functional equations. The preparation of that talk has been the starting point to writing this paper.

Work supported by M.U.R.S.T. Research funds (60%).

REFERENCES

[1] ACZÉL, J., *7. Problem.* In *Report of the twenty-first International Symposium on Functional Equations.* Aequationes Math. *26* (1984), 258.

[2] ALBERT, M. and BAKER, J. A., *Functions with bounded n-th differences.* Ann. Polon. Math. *43* (1983), 93–103.

[3] ALSINA, C., *On the stability of a functional equation related to associativity.* Ann. Polon. Math. *53* (1991), 1–5.

[4] BADORA, R., *On some generalized invariant means and their application to the stability of the Hyers – Ulam type, I.* Manuscript.

[5] BAKER, J. A., *The stability of the cosine equation.* Proc. Amer. Math. Soc. *80* (1980), 411–416.

[6] BAKER, J., LAWRENCE, J. and ZORZITTO, F., *The stability of the equation $f(x + y) = f(x)f(y)$.* Proc. Amer. Math. Soc. *74* (1979), 242–246.

[7] BASTER, J., MOSZNER, Z. and TABOR, J., *On the stability of some class of functional equations.* [Rocznik Naukovo-Dydaktyciny, WSP w Krakowie, Zeszyt 97, Prace Matematyczne XI]. Cracow, 1985.

[8] BEAN, M. and BAKER, J. A., *The stability of a functional analogue of the wave equation.* Canad. Math. Bull. *33* (1990), 376–385.

[9] BORELLI, C., *On Hyers – Ulam stability of Hosszú's functional equation.* Resultate Math. *26* (1994), 221–224.

[10] BORELLI, C. and FORTI, G. L., *On a general Hyers – Ulam stability result.* Internat. J. Math. Math. Sci. *18* (1995), 229–236.

[11] BOURGIN, D. G., *Approximately isometric and multiplicative transformations on continuous function rings.* Duke Math. J. *16* (1949), 385–397.

[12] BOURGIN, D. G., *Classes of transformations and bordering transformations.* Bull. Amer. Math. Soc. *57* (1951), 223–237.

[13] BURKILL, J. C., *Polynomial approximations to functions with bounded differences.* J. London Math. Soc. *33* (1958), 157–161.

[14] CENZER, D., *The stability problem for transformations of the circle.* Proc. Roy. Soc. Edinburgh A *84* (1979), 279–281.

[15] CHMIELINSKI, J. and TABOR, J., *On approximate solutions of the Pexider equation.* Aequationes Math. *46* (1993), 143–163.

[16] CHOLEWA, P. W., *The stability of the sine equation.* Proc. Amer. Math. Soc. *88* (1983), 631–634.

[17] CHOLEWA, P. W., *Remarks on the stability of functional equations.* Aequationes Math. *27* (1984), 76–86.

[18] CHOLEWA, P. W., *The stability problem for a generalized Cauchy type functional equation.* Rev. Roumaine Math. Pures Appl. *29* (1984), 457–460.

[19] CHOLEWA, P. W., *Almost approximately polynomial functions.* In *Nonlinear analysis.* World Sci. Publishing, Singapore, 1987, pp. 127–136.

[20] CZERWIK, St., *On the stability of the quadratic mapping in normed spaces.* Abh. Math. Sem. Univ. Hamburg *62* (1992), 59–64.

[21] DE LA HARPE, P. and KAROUBI, M., *Représentations approchées d'un groupe dans une algèbre de Banach.* Manuscripta Math. *22* (1977), 293–310.

[22] DICKS, D., *2. Remark.* In *Report of the twenty-seventh International Symposium on Functional Equations.* Aequationes Math. *39* (1990), 310.

[23] DJOKOVIĆ, D. Ž., *A representation theorem for $(X_1 - 1)(X_2 - 1) \cdots (X_n - 1)$ and its applications.* Ann. Pol. Math. *22* (1969), 189–198.

[24] DRLJEVIĆ, H., *On the stability of the functional quadratic on A-orthogonal vectors.* Publ. Inst. Math. (Beograd) (N.S.) *36* (50) (1984), 111–118.

[25] DRLJEVIĆ, H., *On the stability of a functional which is approximately additive or approximately quadratic on A-orthogonal vectors.* In *Differential geometry, calculus of variations, and their applications.* [Lecture Notes in Pure and Appl. Math., Vol. 100]. Dekker, New York, 1985, pp. 511–513.

[26] DRLJEVIĆ, H., *The stability of the weakly additive functional.* In *Nonlinear analysis.* World Sci. Publishing, Singapore, 1987, pp. 287–306.

[27] DRLJEVIĆ, H., *On the representation of functionals and the stability of mappings in Hilbert and Banach spaces.* In *Topics in mathematical analysis.* Ser. Pure Math. II, World Sci. Publishing, Singapore, 1989, pp. 231–245.

[28] DRLJEVIĆ, H. and MAVAR, Z., *About the stability of a functional approximately additive on A-orthogonal vectors.* Akad. Nauka Umjet. Bosne Hercegov. Rad. Odjelj. Prirod. Mat. Nauka (1982), No. 20, 155–172.

[29] FENYÖ, I., *Osservazioni su alcuni teoremi di D. H. Hyers.* Istit. Lombardo Accad. Sci. Lett. Rend. A *114* (1980), 235–242 (1982).

[30] FENYÖ, I., *On an inequality of P. W. Cholewa.* In *General inequalities, 5.* [Internat. Schriftenreihe Numer. Math., Vol. 80]. Birkhäuser, Basel–Boston, MA, 1987, pp. 277–280.

[31] FÖRG-ROB, W. and SCHWAIGER, J., *Stability and superstability of the addition and subtraction formulae for trigonometric and hyperbolic functions.* In *Contributions to the Theory of Functional Equations, Proceedings of the Seminar Debrecen-Graz, 1991,* Grazer Math. Ber. *315* (1991), 35–44.

[32] FÖRG-ROB, W. and SCHWAIGER, J., *On the stability of a system of functional equations characterizing generalized hyperbolic and trigonometric functions.* Aequationes Math. *45* (1993), 285–296.

[33] FORTI, G. L., *An existence and stability theorem for a class of functional equations.* Stochastica *4* (1980), 23–30.

[34] FORTI, G. L., *The stability of homomorphisms and amenability, with applications to functional equations.* Abh. Math. Sem. Univ. Hamburg *57* (1987), 215–226.

[35] FORTI, G. L., *Sulla stabilità degli omomorfismi e sue applicazioni alle equazioni funzionali.* Rend. Sem. Mat. Fis. Milano *58* (1988), 9–25 (1990).

[36] FORTI, G. L., *18. Remark.* In *Report of the twenty-seventh International Symposium on Functional Equations.* Aequationes Math. *39* (1990), 309–310.

[37] FORTI, G. L. and SCHWAIGER, J., *Stability of homomorphisms and completeness.* C. R. Math. Rep. Acad. Sci. Canada *11* (1989), 215–220.

[38] FRÉCHET, M., *Les polynomes abstraits.* Journal de Mathématiques *8* (1929), 71–92.

[39] GAJDA, Z., *On stability of the Cauchy equation on semigroups.* Aequationes Math. *36* (1988), 76–79.

[40] GAJDA, Z., *Local stability of the functional equation characterizing polynomial functions.* Ann. Polon. Math. *52* (1990), 119–137.

[41] GAJDA, Z., *On stability of additive mappings.* Internat. J. Math. Math. Sci. *14* (1991), 431–434.
[42] GAJDA, Z., *Note on invariant means for essentially bounded functions.* Manuscript.
[43] GAJDA, Z., *Generalized invariant means and their applications to the stability of homomorphisms.* Manuscript.
[44] GAJDA, Z., *Invariant means and representations of semigroups in the theory of functional equations.* [Prace Naukowe Uniwersytetu Śląskiego w Katowicach, No. 1273]. Uniw. Śląsk., Katowice, 1992.
[45] GAJDA, Z. and GER, R., *Subadditive multifunctions and Hyers – Ulam stability.* In *General inequalities*, 5. [Internat. Schriftenreihe Numer. Math., Vol. 80]. Birkhäuser, Basel–Boston, MA, 1987.
[46] GER, R., *Almost approximately additive mappings.* In *General inequalities, 3.* [Internat. Schriftenreihe Numer. Math., Vol. 64]. Birkhäuser, Basel–Boston, MA, 1983, pp. 263–276.
[47] GER, R., *Stability of addition formulae for trigonometric mappings.* Zeszyty Naukowe Politechniki Slasiej, Ser. Matematyka-Fizyka z. *64* (1990), no. 1070, 75–84.
[48] GER, R., *On functional inequalities stemming from stability questions.* In *General inequalities, 6.* [Internat. Schriftenreihe Numer. Math., Vol. 103]. Birkhäuser, Basel–Boston, Mass., 1992.
[49] GER, R., *Superstability is not natural.* In *Report of the twenty-sixth International Symposium on Functional Equations.* Aequationes Math. *37* (1989), 68.
[50] GER, R., *The singular case in the stability behaviour of linear mappings.* Grazer Math. Ber. *316* (1991), 59–70.
[51] GREENLEAF, F. P., *Invariant means on topological groups.* [Van Nostrand Mathematical Studies, Vol. 16]. Van Nostrand, New York–Toronto–London–Melbourne, 1969.
[52] GROTHENDIECK, A., *Produits tensoriels topologiques et espace nucléaires.* [Memoirs Amer. Math. Soc., No. 16]. A.M.S., Providence, R.I., 1955.
[53] HYERS, D. H., *On the stability of the linear functional equation.* Proc. Nat. Acad. Sci. U.S.A. *27* (1941), 222–224.
[54] HYERS, D. H., *Transformations with bounded n-th differences.* Pacific J. Math. *11* (1961), 591–602.
[55] HYERS, D. H., *The stability of homomorphisms and related topics.* In *Global analysis — analysis on manifolds.* [Teubner-Texte Math. 57]. Teubner, Leipzig, 1983, pp. 140–153.
[56] HYERS, D. H. and RASSIAS, TH. M., *Approximate homomorphisms.* Aequationes Math. *44* (1992), 125–153.
[57] ISAC, G. and RASSIAS, TH. M., *On the Hyers – Ulam stability of ψ-additive mappings.* J. Approx. Theory *72* (1993), 131–137.
[58] JAROSZ, K., *Perturbations of Banach algebras.* [Lecture Notes in Mathematics, Vol. 1120]. Springer-Verlag, Berlin–Heidelberg–New York–Tokyo, 1985.
[59] JOHNSON, B. E., *Cohomology in Banach algebras.* [Memoirs Amer. Math. Soc., No. 127]. A.M.S., Providence, R.I., 1972.
[60] JOHNSON, B. E., *Approximately multiplicative functionals.* J. London Math. Soc. (2) *34* (1986), 489–510.
[61] JOHNSON, B. E., *Approximately multiplicative maps between Banach algebras.* J. London Math. Soc. (2) *37* (1988), 294–316.
[62] KAZHDAN, D., *On ε-representations.* Israel J. Math. 43 (1982), 315–323.
[63] KOMINEK, Z., *On a local stability of the Jensen functional equation.* Demonstratio Math. *22* (1989), 499–507.
[64] LAWRENCE, J., *The stability of multiplicative semigroup homomorphisms to real normed algebras.* I. Aequationes Math. *28* (1985), 94–101.
[65] MAZUR, S. and ORLICZ, W., *Grundelegende Eigenschaften der polynomischen Operationen. Erste Mitteilung.* Studia Math. *5* (1934), 50–68.
[66] MAZUR, S. and ORLICZ, W., *Grundelegende Eigenschaften der polynomischen Operationen. Zweite Mitteilung.* Studia Math. *5* (1934), 179–189.
[67] MOSZNER, Z., *Sur la stabilité de l'équation d'homomorphisme.* Aequationes Math. *29* (1985), 290–306.
[68] MOSZNER, Z., *Sur la définition de Hyers de la stabilité de l'équation fonctionnelle.* Opuscula Math. (1987), 47–57 (1988).

[69] NASHED, M. Z. and VOTRUBA, G. F., *A unified operator theory of generalized inverses.* In *Generalized inverses and applications.* Academic Press, New York–San Francisco–London, 1976, pp. 1–109.

[70] NIKODEM, K., *The stability of the Pexider equation.* Ann. Math. Sil. *5* (1991), 91–93.

[71] PAGANONI, L., *Soluzione di una equazione funzionale su dominio ristretto.* Boll. Un. Mat. Ital. (5) 17-B (1980), 979–993.

[72] PÓLYA, G. and SZEGÖ, G., *Problems and theorems in analysis,* Vol. I, Part One, Ch. 3, Problem 99. [Grundlehren der mathematischen Wissenschaften in Einzeldarstellungen, Band 193]. Springer-Verlag, Berlin–Heidelberg–New York, 1972.

[73] RASSIAS, J. M., *On approximation of approximately linear mappings by linear mappings.* J. Funct. Anal. *46* (1982), 126–130.

[74] RASSIAS, J. M., *On approximation of approximately linear mappings by linear mappings.* Bull. Sci. Math. (2) *108* (1984), 445–446.

[75] RASSIAS, J. M., *On a new approximation of approximately linear mappings by linear mappings.* Discuss. Math. *7* (1985), 193–196.

[76] RASSIAS, J. M., *Solution of a problem of Ulam.* J. Approx. Theory *57* (1989), 268–273.

[77] RASSIAS, J. M., *On the stability of the Euler–Lagrange functional equation.* C. R. Acad. Bulgare Sci. *45* (1992), 17–20.

[78] RASSIAS, TH. M., *On the stability of the linear mapping in Banach spaces.* Proc. Amer. Math. Soc. *72* (1978), 297–300.

[79] RASSIAS, TH. M., *The stability of linear mappings and some problems on isometrices.* In *Mathematical analysis and its applications* (Kuwait, 1985). [KFAS Proc. Ser., Vol. 3]. Pergamon, Oxford-Elmsford, NY, 1988, pp. 175–184.

[80] RASSIAS, TH. M., *On the stability of mappings.* Rend. Sem. Mat. Fis. Milano *58* (1988), 91–99 (1990).

[81] RASSIAS, TH. M., *On a modified Hyers–Ulam sequence.* J. Math. Anal. Appl. *158* (1991), 106–113.

[82] RASSIAS, TH. M. and ŠEMRL, P., *On the behaviour of mappings which do not satisfy Hyers–Ulam stability.* Proc. Amer. Math. Soc. *114* (1992), 989–993.

[83] RASSIAS, TH. M. and ŠEMRL, P., *On the Hyers–Ulam stability of linear mappings.* J. Math. Anal. Appl. *173* (1993), 325–338.

[84] RÄTZ, J., *On approximately additive mappings.* In *General inequalities,* 2. [Internat. Schriftenreihe Numer. Math., Vol. 47]. Birkhäuser, Basel–Boston, MA, 1980, pp. 233–251.

[85] SCOTT, W. R., *Group theory.* Dover Publications, New York, 1987.

[86] ŠEMRL, P., *The stability of approximately additive functions.* Manuscript.

[87] ŠEMRL, P., *Isomorphisms of standard operator algebras.* To appear in Proc. Amer. Math. Soc.

[88] ŠEMRL, P., *The functional equation of multiplicative derivation is superstable on standard operator algebras.* Integral Equations Operator Theory *18* (1994), 118–122.

[89] SCHWAIGER, J., *On the stability of a functional equation for homogeneous functions.* In Report of the twenty-second International Symposium on Functional Equations. Aequationes Math. *29* (1985), 80.

[90] SHAPIRO, H. N., *Note on a problem in number theory.* Bull. Amer. Math. Soc. *54* (1948), 890–893.

[91] SHTERN, A. I., *On stability of homomorphisms in the group ℝ*.* Vestnik MGU Ser. Matem. Mech. *37* (1982), 29–32. — English translation in Moscow Univ. Math. Bull. *37* (1982), 33–36.

[92] SHTERN, A. I., *Quasirepresentations and pseudorepresentations.* Funktsional. Anal. Prilozhen. *25* (1991), 70–73. — English translation in Funct. Anal. Appl. 25 (1991), 140–143.

[93] SKOF, F., *Sull'approssimazione delle applicazioni localmente δ-additive.* Atti Accad. Sci. Torino Cl. Sci. Fis. Mat. Natur. *117* (1983), 377–389 (1986).

[94] SKOF, F., *Proprietà locali e approssimazione di operatori.* In *Geometry of Banach spaces and related topics (Milan, 1983).* Rend. Sem. Mat. Fis. Milano *53* (1983), 113–129 (1986).

[95] SKOF, F., *Approssimazione di funzioni δ-quadratiche su dominio ristretto.* Atti Accad. Sci. Torino Cl. Sci. Fis. Mat. Natur. *118* (1984), 58–70.

[96] SKOF, F., *On approximately quadratic functions on a restricted domain*. In *Report of the third International Symposium on Functional Equations and Inequalities, 1986*. Publ. Math. Debrecen *38* (1991), 14.

[97] SMAJDOR, A., *Hyers–Ulam stability for set valued functions*. In *Report of the twenty-seventh International Symposium on Functional Equations*. Aequationes Math. *39* (1990), 297.

[98] SZÉKELYHIDI, L., *The stability of linear functional equations*. C. R. Math. Rep. Acad. Sci. Canada *3* (1981), 63–67.

[99] SZÉKELYHIDI, L., *On a stability theorem*. C. R. Math. Rep. Acad. Sci. Canada 3 (1981), 253–255.

[100] SZÉKELYHIDI, L., *On a theorem of Baker, Lawrence and Zorzitto*. Proc. Amer. Math. Soc. *84* (1982), 95–96.

[101] SZÉKELYHIDI, L., *The stability of d'Alembert-type functional equations*. Acta Sci. Math. (Szeged) *44* (1982), 313–320 (1983).

[102] SZÉKELYHIDI, L., *Note on a stability theorem*. Canad. Math. Bull. *25* (1982), 500–501.

[103] SZÉKELYHIDI, L., *Note on Hyers's theorem*. C. R. Math. Rep. Acad. Sci. Canada ·*8* (1986), 127–129.

[104] SZÉKELYHIDI, L., *Remarks on Hyers's theorem*. Publ. Math. Debrecen *34* (1987), 131–135.

[105] SZÉKELYHIDI, L., *Fréchet's equation and Hyers theorem on noncommutative semigroups*. Ann. Polon. Math. *48* (1988), 183–189.

[106] SZÉKELYHIDI, L., *Stability of some functional equations in economics*. Rend. Sem. Mat. Fis. Milano *58* (1988), 169–176 (1990).

[107] SZÉKELYHIDI, L., *An abstract superstability theorem.* Abh. Math. Sem. Univ. Hamburg *59* (1989), 81–83.

[108] SZÉKELYHIDI, L., *Stability properties of functional equations describing the scientific laws*. J. Math. Anal. Appl. *150* (1990), 151–158.

[109] SZÉKELYHIDI, L., *The stability of the sine and cosine functional equations*. Proc. Amer. Math. Soc. *110* (1990), 109–115.

[110] SZÉKELYHIDI, L., *Stability properties of functional equations in several variables*. Manuscript.

[111] TABOR, J., *Ideal stability of the Cauchy and Pexider equations*. In Report of the twenty-second International Symposium on Functional Equations, Aequationes Math. *29* (1985), 82.

[112] TABOR, J., *Ideal stability of the Cauchy equation*. In *Proceedings of the twenty-third International Symposium on Functional Equations*. Centre for Inf. Th., Univ. of Waterloo, Waterloo, Ont., 1985.

[113] TABOR, J., *On functions behaving like additive functions*. Aequationes Math. *35* (1988), 164–185.

[114] TABOR, J., *Quasi-additive functions*. Aequationes Math. *39* (1990), 179–197.

[115] TABOR, J., *Approximate endomorphisms of the complex field*. J. Natur. Geom. *1* (1992), 71–86.

[116] ULAM, S. M., *A collection of mathematical problems*. Interscience Publ., New York, 1961. *Problems in Modern Mathematics*, Wiley, New York, 1964.

[117] ULAM, S. M., *Sets, numbers, and universes*. M.I.T. Press, Cambridge, 1974.

[118] WHITNEY, H., *On functions with bounded n-th differences*. J. Math. Pures Appl. (9) *36* (1957), 67–95.

[119] WHITNEY, H., *On bounded functions with bounded n-th differences*. Proc. Amer. Math. Soc. *10* (1959), 480–481.

[120] ZORZITTO, F., *31. Problem*. In *Report of the twenty-sixth International Symposium on Functional Equations*. Aequationes Math. *37* (1989), 118.

Dipartimento di Matematica,
Università degli Studi di Milano,
Via C. Saldini 50,
I-20133 Milano,
Italia.

Aequationes Mathematicae **50** (1995) 191–213
University of Waterloo

0001–9054/95/020191–23 $1.50 + 0.20/0
© 1995 Birkhäuser Verlag, Basel

Rep-tiling Euclidean space

ANDREW VINCE

Summary. A *rep-tiling* $\mathcal{T}$ is a self replicating, lattice tiling of R^n. *Lattice tiling* means a tiling by translates of a single compact tile by the points of a lattice, and *self-replicating* means that there is a non-singular linear map $\phi: R^n \to R^n$ such that, for each $T \in \mathcal{T}$, the image $\phi(T)$ is, in turn, tiled by $\mathcal{T}$. This topic has recently come under investigation, not only because of its recreational appeal, but because of its application to the theory of wavelets and to computer addressing. The paper presents an exposition of some recent results on rep-tiling, including a construction of essentially all rep-tilings of Euclidean space. The construction is based on radix representation of points of a lattice. One particular radix representation, called the *generalized balanced ternary*, is singled out as an example because of its relevance to the field of computer vision.

1. Introduction

The subject of this exposition, self-replicating tiling, has gained the interest of a wide spectrum of mathematicians. It is a recent addition to the large body of work on the geometry and symmetry of tilings, a topic surveyed, beginning with the mosaics in the Alhambra at Granada in Spain, in the book [15] by Grünbaum and Shephard. Self-replicating tiling also relates to fractal geometry. The boundaries of the tiles often have nonintegral Hausdorff dimension, and techniques have been developed for computing the dimension. Self-replicating tilings are connected with generalized number systems, a topic that dates back at least to Cauchy, who noted that allowing negative digits in the radix representation of an integer makes it unnecessary for a person to memorize the multiplication table past 5×5. Knuth [19] discusses numerous alternative positional number systems, in particular the balanced ternary system, whose base is 3 and whose digits are the "trits" $\{-1, 0, 1\}$. Self-replicating tilings arise in image processing and computer vision, especially in the addressing of points in the plane using a hexagonal, rather than a square, grid

---

AMS (1991) subject classification: Primary 52C20, 52C22, Secondary 11A63.

Manuscript received February 26, 1993 and, in final form, September 26, 1994.

of pixels. In this case a system generalizing the balanced ternary system comes into play. Self-replicating tilings have recently been applied to the construction of wavelets. The standard wavelet bases are constructed using translations and expansions from simple functions with support on one tile in the usual cubic tiling of $\mathbb{R}^n$. Gröchenig and Madych [13], Lawton and Resnikoff [23] and Strichartz [31] use multiresolution analysis modeled on other self-replicating tilings of $\mathbb{R}^n$. In 1984 Shechtman, Blech, Gratias and Cahn [30] discovered the first substance (an aluminum-manganese alloy) whose electron diffraction pattern indicates both "long range order" and violation of the crystallographic restriction (five-fold rotational symmetry in this case). Long range order usually means periodicity, but periodicity is incompatible with five-fold symmetry. Although the arrangement of atoms in this and similar materials, now called *quasicrystals*, is still unknown, certain self-replicating tilings due to Penrose [27] and others have become a canonical model for their structure. Thurston [32] makes basic connections between self-replicating tilings, finite state machines and Markov partitions in dynamical systems. The "expansion function" of a self-replicating lattice tiling of $\mathbb{R}^n$ induces a self map of the torus, the torus being the quotient of $\mathbb{R}^n$ by the lattice group of isometries isomorphic to $\mathbb{Z}^n$. Radin [28], in attempting to find the extent of disorder possible in certain tilings, uses the expansion function of a self-replicating tiling to construct another dynamical system on a certain space of tilings. He makes connections between the symmetry of the tilings and ergodic theory and statistical mechanics.

The intent of this paper is not an exhaustive survey of the topics mentioned above, but an introductory exposition of the subject for an interested nonspecialist. After giving a definition of rep-tiling in Section 2, a correspondence between rep-tiles and radix systems is presented in Section 3. As a trivial example, the standard base 10 radix system corresponds to the tiling of the real line by unit intervals with the following self-replicating property: an expansion of each tile by a factor of ten results in a tiling of the line by (first level) tiles, each of which is the union of ten of the original (zero level) tiles. Continuing this process leads to the hierarchy upon which ordinary arithmetic is based: for each $m \geq 1$, the line is tiled by mth level tiles, each of which is tiled, in turn, by $(m-1)$st level tiles. The main result in Section 3 is a bijection between pure lattice rep-tilings of $\mathbb{R}^n$ and n-dimensional radix systems satisfying unique representation. Section 4 deals with a particular radix system relevant to computer vision. The corresponding tiling of space, in this case, is by permutohedra (hexagons in dimension 2, truncated octahedra in dimension 3, ...), and the radix system is a generalization of the balanced ternary. No necessary and sufficient conditions for unique representation are known, but Section 5 provides several sufficient conditions for a radix system to possess the desired unique representation property. The term "self-replicating", as used in the first paragraph of this paper, is generic in the sense that it has slightly different definitions depending on the context. In particular, concern in this paper is mainly with lattice tilings; generalizations, variations and open problems will be discussed briefly in Section 6. For most proofs, the reader willl be referred to the appropriate source.

2. Rep-tiling

In a tiling, all tiles are compact; the tiles cover $\mathbb{R}^n$; and the intersection of the interiors of any two distinct tiles is empty. Most tilings in this paper are lattice tilings with a certain self-replicating property. More precisely, a *lattice* in $\mathbb{R}^n$ is the set of all integer combinations of n linearly independent vectors, and a *lattice tiling* is a tiling $\mathscr{T}$ of $\mathbb{R}^n$ by translates of a single tile T by a lattice L. In other words, $\mathscr{T} = \{x + T \,|\, x \in L\}$. The common wall tilings by squares or by hexagons are examples of lattice tilings.

The self-replicating property goes back at least to 1964 when Golomb [12] defined a figure F to be *rep-k* if F can be tiled by k congruent similar figures. Three rep-4 figures are shown in Figure 1. Combining the notion of rep-k figure with the notion of tiling, Figure 2 shows three tilings of the plane where the tiles are the corresponding rep-4 figures in Figure 1. Each of the examples in Figure 2 is a tiling $\mathscr{T}$ having the property that there exists a similarity, i.e. a matrix of the form $A = bQ$ with Q orthogonal and b positive real, such that for each tile T the image $A(T)$ is, in turn, tiled by copies of tiles in $\mathscr{T}$. In the first and second examples the similarity is expansion by a factor of 2; in the third example the similarity is a $\pi/2$ rotation composed with expansion by a factor of 2. The first example is a lattice tiling, but the other two examples are not. In fact, the third example is not even *periodic*, which means there do not exist translations in two linearly independent directions that preserve the tiling.

A rep-tiling is slightly more general than the examples above in that arbitrary expansive matrices, not just similarities, are allowed. A matrix is called *expansive* if all eigenvalues have modulus greater than 1. If $A = bQ$ is a similarity, then A expansive is equivalent to $b > 1$. A *rep-tiling* is a tiling $\mathscr{T}$ by translates of a single

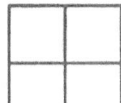  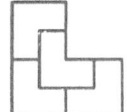

Figure 1. Rep-4 figures.

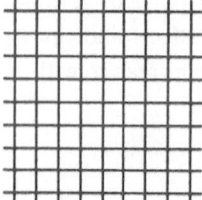 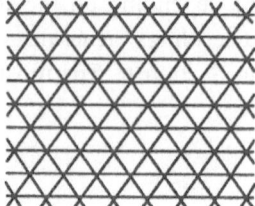

Figure 2. Rep-4 tilings.

tile T_0 such that

(1) T_0 is compact with nonempty interior, and
(2) there exists an expansive matrix A such that for each tile T the image $A(T)$ is, in turn, tiled by copies of tiles in $\mathcal{T}$.

It is sufficient to assume in condition (1) that T_0 has positive Lebesque measure (instead of nonempty interior), and it follows from the definition that T_0 is actually the closure of its interior and that the boundary of T_0 has Lebesque measure zero [20]. Each tile in a rep-tiling is a rep-k figure. It follows from the definition and the fact that distinct pairs of tiles intersect in a set of measure zero that $k = |\det A|$. In particular, $\det A$ must be an integer. If, in addition to conditions (1) and (2),

(3) $\mathcal{T}$ is a lattice tiling,

then $\mathcal{T}$ is called a *lattice rep-tiling*. The first example in Figure 2 is a lattice rep-tiling, but the other two rep-tilings are not lattice rep-tilings. In fact, the only example of a lattice rep-tiling given so far is the standard tiling of the plane by squares. What seems surprising at first is that there are infinitely many lattice rep-tilings in each dimension. A construction is given in Section 3. Lattice rep-tilings have been investigated independently by Kenyon [17], Gröchenig and Hass

[14], Gröchenig and Madych [13], Bandt [1] and Vince [34], and more recently by Lagarias and Wang [20]–[22] and Gelbrich [5].

3. Radix representation

A basic result in number theory states that every non-negative integer has a unique base $\beta \geq 2$ representation of the form

$$\sum_{i=0}^{m} d_i \beta^i, \tag{1}$$

where $d_i \in D = \{0, 1, \ldots, \beta - 1\}$. Here D is called the *digit set* and β is called the *radix*. The representation (1) has been generalized in several ways. In each case a central issue is unique representation.

(i) The radix need not be positive. In fact, for $\beta \leq -2$ every integer, including the negatives, has a base β representation with digit set $D = \{0, 1, \ldots, |\beta| - 1\}$. The digits can also be negative. Particularly nice is the *balanced ternary* system, where the radix is 3 and the digit set is $D = \{-1, 0, 1\}$. Every integer has a unique radix representation in the balanced ternary system.

(ii) In 1981 Gilbert [7]–[9] extended radix representation to the Gaussian integers $\mathbb{Z}[i] = \{a + bi \,|\, a, b \in \mathbb{Z}\}$. For example, every Gaussian integer has a unique radix $\beta = -1 + i$ representation of the form (1), where $d_i \in D = \{0, 1\}$ — hence a binary system for the Gaussian integers. The radix β arithmetic in the Gaussian integers resembles usual binary arithmetic except in the carry digits. For example, $1 + 1 = 1100$ because $2 = \beta^3 + \beta^2$. So $1 + 1$ results in 0 with 110 "carried" three places to the left. Surprisingly, with $\beta = 1 + i$ instead of $-1 + i$, not every Gaussian integer has a representation; for example i does not.

(iii) Radix representation can likewise be extended to other number fields. Let $f(x) \in \mathbb{Z}[x]$ be a monic polynomial, irreducible over $\mathbb{Z}$, α a root of $f(x)$ in some extension field of the rationals, $\mathbb{Z}[\alpha]$ the ring obtained by adjoining α to $\mathbb{Z}$. In the case $f(x) = x^2 + 1$, the ring $\mathbb{Z}[\alpha]$ is the Gaussian integers given in (ii) above. Another interesting example is $f(x) = x^2 + x + 1$. In this case $\alpha = -\frac{1}{2} + (\sqrt{3}/2)i$, and $\mathbb{Z}[\alpha]$ can be viewed geometrically as the hexagonal lattice in the complex plane (the lattice points being the centers of the hexagons in the hexagonal tiling). If the radix is chosen as $\beta = \frac{5}{2} + (\sqrt{3}/2)i$ and the digit set is $D = \{0, 1, \omega, \omega^2, \ldots, \omega^5\}$, where $\omega = \frac{1}{2} + (\sqrt{3}/2)i$ (so that D consists of 0 and the sixth roots of unity), then every point of $\mathbb{Z}[\alpha]$ has a unique radix representation for the form (1) [18]. It will become apparent in Section 4 that this hexagonal system is a natural 2-dimensional generalization of the balanced ternary, the digit set D playing the role of the "trits".

Consider the following general framework for all the examples above. Let L be a lattice, viewed either geometrically as a set of points in Euclidean space, or algebraically as a finitely generated free Abelian group. Let $A : L \to L$ be a group endomorphism, and D a finite subset of L containing 0. The map A can, without loss of generality, be regarded as any square nonsingular matrix, as long as L is A-invariant. Indeed, if the basis for this matrix is chosen in L, then A is an integer matrix. The triple (L, A, D) is said to have the *unique representation property* if every element of L has a unique finite representation of the form

$$\sum_{i=0}^{m} A^i(d_i),$$ (2)

where $d_i \in D$. If L has a ring structure and A is the matrix that represents multiplication by an element β in the ring (the endomorphism $x \mapsto \beta x \ \forall x \in L$), then expression (2) reduces to the form of expression (1). In this lattice framework the triples (L, A, D) corresponding to the examples above — the balanced ternary, the Gaussian integers with base $-1 + i$ and the hexagonal example — are, respectively:

(i) $L = \mathbb{Z}, \quad A = (3), \quad D = \{-1, 0, 1\};$

(ii) $L = \mathbb{Z}[i], \quad A = \begin{pmatrix} -1 & -1 \\ 1 & 1 \end{pmatrix}, \quad D = \{0, 1\};$

(iii) $L = \mathbb{Z}\left[-\frac{1}{2} + \frac{\sqrt{3}}{2} i \right], \quad A = \begin{pmatrix} \frac{5}{2} & -\frac{\sqrt{3}}{2} \\ \frac{\sqrt{3}}{2} & \frac{5}{2} \end{pmatrix}, \quad D = \{0, 1, \omega, \omega^2, \ldots, \omega^5\}.$

In the last two examples, the matrix A is with respect to the standard basis. In the rest of this paper the examples above, all of which possess the unique representation property, will be referred to as *Examples 1, 2 and 3*.

The following proposition gives two necessary conditions for unique representation in (L, A, D) and a sufficient condition for uniqueness. The reader is referred to [34] for the somewhat technical proof of the second statement. A *digit set* for (L, A) is a complete set of residues for L modulo $A(L)$; in other words D is a digit set if it contains exactly one representative from each coset in the quotient group $L/A(L)$. The number of digits, i.e. the number of cosets, is, by standard algebraic techniques, equal to $|\det A|$.

PROPOSITION. (1) *If* (L, A, D) *has the unique representation property then D is a digit set. If D is a digit set then representation of a point, if it exists, is unique.*

(2) *If* (L, A, D) *has the unique representation property then A is an expansive map.*

Proof of (1). Assume that (L, A, D) has the unique representation property. To show that no coset is represented twice, assume, by way of contradiction, that $d \equiv d'$ mod $A(L)$ for some $d, d' \in D, d \neq d'$. This implies that $d' = d + A(x) = d + A(\sum_{i=0}^{m} A^i(d_i))$ for some $x \in L$ and some $d_i \in D$. Then $d' = d + \sum_{i=1}^{m+1} A^i(d_i)$, which contradicts uniqueness. To show that each coset has at least one representative in D, consider any $x \in L$. Then $x = \sum_{i=0}^{m} A^i(d_i) = d_0 + \sum_{i=1}^{m} A^i(d_i)$ for some $d_i \in D$ implies that $x \equiv d_0$ mod $A(L)$.

Concerning the second statement, let D be a digit set and assume uniqueness of representation is violated. Then $\sum_{i=0}^{m} A^i(d_i) = \sum_{i=0}^{m} A^i(d'_i)$ for some $d_i, d'_i \in D$, and since A is invertible it may be assumed, without loss of generality, that $d_0 \neq d'_0$. But then $d_0 \equiv d'_0$ mod $A(L)$, a contradiction. $\square$

A triple (L, A, D) will be called a *radix system* if A is expansive and D is a digit set. Hence if (L, A, D) is a radix system, unique representation is, by part (1) of the proposition, reduced to showing that each lattice point has some representation. The two necessary conditions in the proposition, however, are not sufficient to insure that each lattice point has a representation, even in dimension 1. For example, with 3 as radix, $D = \{-1, 0, 4\}$ is a complete set of residues modulo 3, i.e. a digit set. However, -2 has no radix 3 representation. Conditions under which each integer has a unique radix representation have been investigated by Matula [25] and by Odlyzko [26]. There seems to be no known simple necessary and sufficient conditions to insure representation. We will return to this problem in Section 5.

Given a radix system (L, A, D), a set $T(A, D)$ is constructed as follows, where the sum is the Minkowski sum $\sum_{i=0}^{\infty} X_i = \{x_0 + x_1 + \cdots \mid x_i \in X_i\}$ in $\mathbb{R}^n$:

$$T(A, D) = \sum_{i=1}^{\infty} A^{-i}(D). \tag{3}$$

Translating $T(A, D)$ by the lattice L gives

$$\mathcal{T}(L, A, D) = \{x + T \mid x \in L\}. \tag{4}$$

It is not obvious that $\mathcal{T}(L, A, D)$ is a tiling of $\mathbb{R}^n$ or even that $T(A, D)$ is a tile (compact with nonempty interior). This turns out to be the case, however, when

(L, A, D) is a radix system. Moreover, the following theorem relates rep-tiling to unique representation. A rep-tiling is called *pure* if the origin lies in the interior of some tile.

THEOREM 1. (1) *If (L, A, D) is a radix system having the unique representation property, then $\mathcal{T}(L, A, D)$ is a pure lattice rep-tiling.*

(2) *If $\mathcal{T}$ is a pure lattice rep-tiling, then $\mathcal{T} = \mathcal{T}(L, A, D)$ for some radix system (L, A, D) having the unique representation property.*

Using a computer implementation of formulas (3) and (4), the tiling in Figure 3 is constructed from Gilbert's binary system for the Gaussian integers (Example 2) and produces a rep-tiling by the rep-2 "dragon" tile. The tiling in Figure 4 is constructed from the hexagonal Example 3 and produces a rep-tiling by the rep-7 "flowsnake." A rep-5 tiling appears in Figure 5. The three tiles (rep-4 Sierpinski triangle, rep-3 figure and rep-9 figure) appearing in Figures 6–8 also produce a rep-tiling by translation of the tile by the respective lattice. These examples show that individual tiles may be topologically complex, not necessarily connected or simply connected. In fact, the tile in Figure 8 has infinitely many connected components.

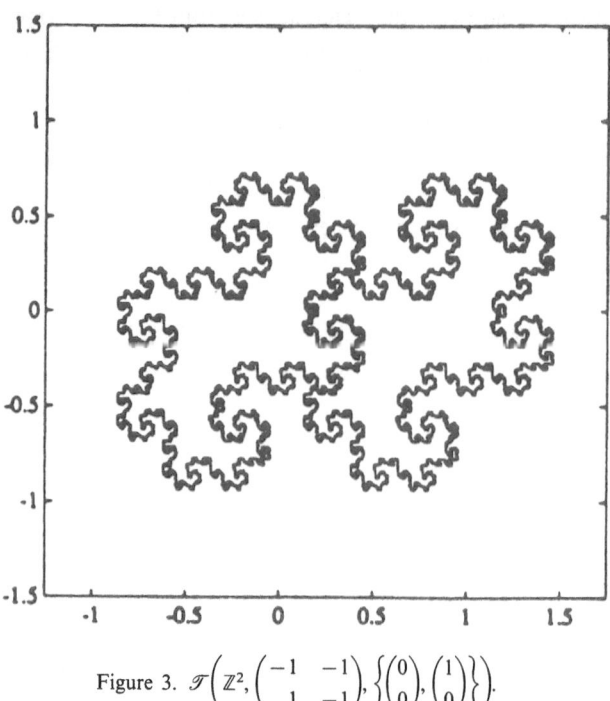

Figure 3. $\mathcal{T}\left(\mathbb{Z}^2, \begin{pmatrix} -1 & -1 \\ 1 & -1 \end{pmatrix}, \left\{ \begin{pmatrix} 0 \\ 0 \end{pmatrix}, \begin{pmatrix} 1 \\ 0 \end{pmatrix} \right\} \right).$

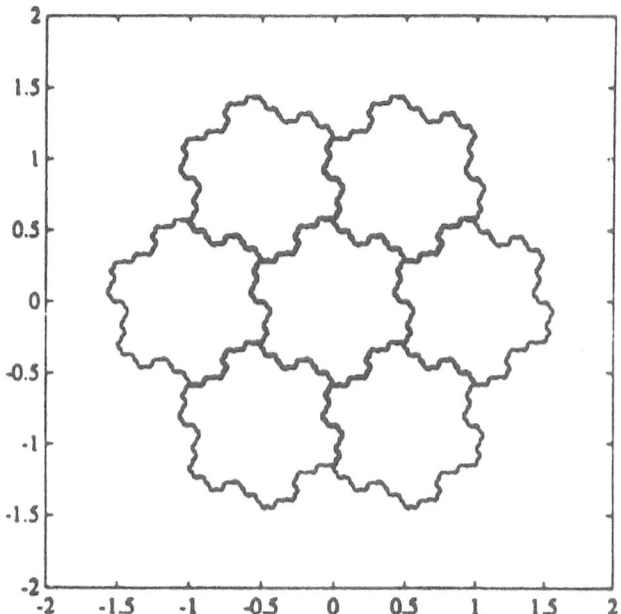

Figure 4. $\mathscr{T}\left(L, \begin{pmatrix} 5/2 & -\sqrt{3}/2 \\ \sqrt{3}/2 & 5/2 \end{pmatrix}, \{(0, 1, \omega, \ldots, \omega^5)\}\right)$, where L is the hexagonal lattice.

It is not entirely surprising that the boundary of the tiles are fractal. Mandelbrot [24], Giles [10] and Gilbert [7] had constructed rep-k fractal figures (but not tilings). Dekking [3]–[4] and Bandt [1] subsequently gave systematic constructions for such "fractiles". Dekking constructs the boundary of a rep-tile in R^2 by a recursive string-rewriting procedure, and Bandt constructs the tile using summation (3).

Omitting the technicalities of the proof of Theorem 1, it is, nevertheless, not difficult to see, given the rep-tiling, how the digits in the radix representation arise, and conversely, given the digits, how formulas (3) and (4) for the tiling arise: The definition of lattice rep-tiling in Section 2 implies that for any such tiling there is a set $D = \{d_1, d_2, \ldots, d_k\}$ consisting of $k = |\det A|$ lattice points such that if T_0 is the tile whose interior contains the origin, then

$$A(T_0) = \bigcup_{k=1}^{k} (d_i + T_0), \tag{5}$$

where the digit d_1 can be chosen to be the origin 0. The set D is the required digit set.

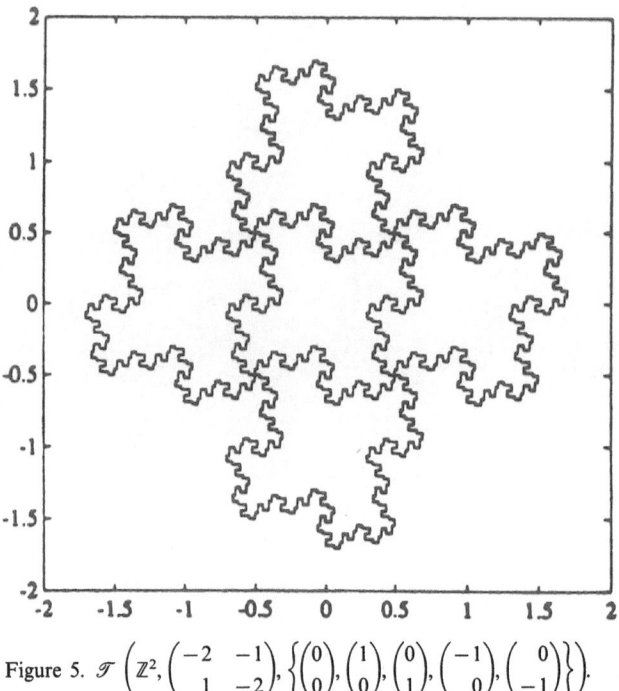

Figure 5. $\mathcal{T}\left(\mathbb{Z}^2, \begin{pmatrix} -2 & -1 \\ 1 & -2 \end{pmatrix}, \left\{ \begin{pmatrix} 0 \\ 0 \end{pmatrix}, \begin{pmatrix} 1 \\ 0 \end{pmatrix}, \begin{pmatrix} 0 \\ 1 \end{pmatrix}, \begin{pmatrix} -1 \\ 0 \end{pmatrix}, \begin{pmatrix} 0 \\ -1 \end{pmatrix} \right\} \right)$.

Conversely, given the radix system (L, A, D), let

$$D_m = \sum_{i=0}^{m-1} A^i(D)$$

denote the set of all lattice points that can be represented with at most m digits. It is equivalent to formula (3) to express $T(A, D)$ as the limit (in the Hausdorff metric) of a nested sequence of sets:

$$T(A, D) = \lim_{m \to \infty} A^{-m}(D_m). \tag{6}$$

For a similarity A, the "evolution" of the set $T(A, D)$ in the limit (6) can be nicely visualized. Recall that the Voronoi cell centered at the lattice point $x \in L$ is defined as the set of points y such that y is at least as close to x as to any other point of the lattice: $V_x = \{y \in \mathbb{R}^n : |y - x| \le |y - z| \text{ for all } z \in L\}$. Let $\overline{D_m} = \bigcup_{x \in D_m} V_x$ be the union of the Voronoi cells V_x centered at the points in D_m. Then, by equation (6),

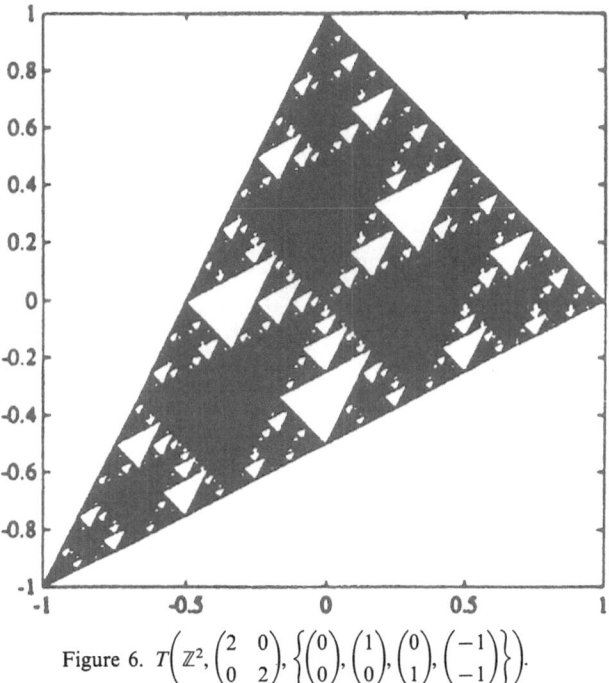

Figure 6. $T\left(\mathbb{Z}^2, \begin{pmatrix} 2 & 0 \\ 0 & 2 \end{pmatrix}, \left\{\begin{pmatrix} 0 \\ 0 \end{pmatrix}, \begin{pmatrix} 1 \\ 0 \end{pmatrix}, \begin{pmatrix} 0 \\ 1 \end{pmatrix}, \begin{pmatrix} -1 \\ -1 \end{pmatrix}\right\}\right)$.

$\overline{D_m}$ (scaled down by A^{-m}) is an mth approximation to T. Figure 9 shows the first approximations to the "dragon tile" $T(A, D)$ of Figure 3.

Since D is a set of coset representatives of $L/A(L)$, the set D_m is a set of coset representatives of the group $L/A^m(L)$. Therefore L is the disjoint union

$$L = \bigcup \{x + D_m \mid x \in A^m(L)\}, \tag{7}$$

which implies that

$$A^{-m}(L) = \bigcup \{x + A^{-m}(D_m) \mid x \in L\}.$$

Letting $m \to \infty$ shows that $\mathbb{R}^n$ is, indeed, covered by copies of the tile $T(A, D)$. (What is not clear is that the interiors of distinct tiles are disjoint, and, in fact, this may not be the case if (L, A, D) does not satisfy the unique representation property. We return to this question in Section 6.) In the case that (L, A, D) does satisfy the unique representation property, let $T_0 = T(A, D)$ and $T_m = \bigcup \{x + T(A, D) \mid x \in D_m\}$ for $m \geq 1$. Then by equation (7)

$$\mathcal{T}_m = \{x + T_m \mid x \in A^m(L)\}$$

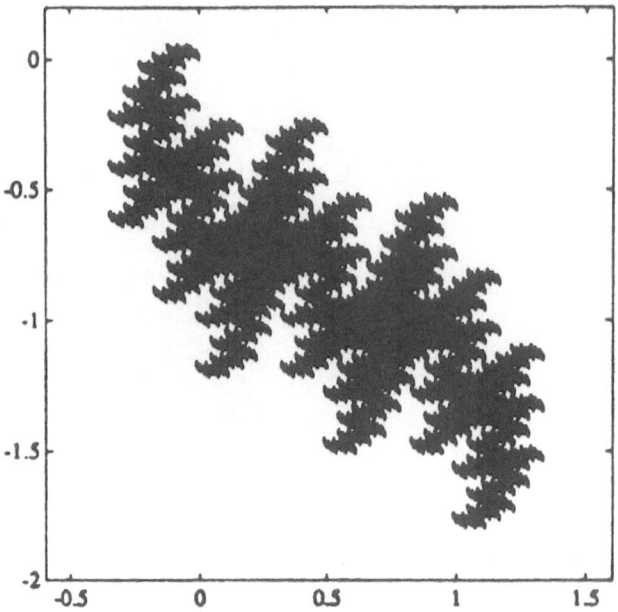

Figure 7. $T\left(L, \begin{pmatrix} 3/2 & -\sqrt{3}/2 \\ \sqrt{3}/2 & 3/2 \end{pmatrix}, \left\{ \begin{pmatrix} 0 \\ 0 \end{pmatrix}, \begin{pmatrix} 1 \\ 0 \end{pmatrix}, \begin{pmatrix} 2 \\ 0 \end{pmatrix} \right\} \right)$, where L is the hexagonal lattice.

is a tiling of $\mathbb{R}^n$ for each m with the following hierarchical property: each tile in $\mathcal{T}_m$ is, in turn, tiled by $|\det A|$ tiles in $\mathcal{T}_{m-1}$.

There is another way to view the tiling; $T(A, D)$ arises as the attractor of a certain iterated function system. More precisely, the functions

$$w_i(x) = A^{-1}(d_i + x),$$

$i = 1, 2, \ldots, k$, are, in the terminology of Barnsley [2], an affine iterated function system, and its *attractor* is, by definition, the unique fixed point of the transformation W defined on the space of all compact subsets of $\mathbb{R}^n$ by

$$W(X) = \bigcup_{i=1}^{k} w_i(X) = \bigcup_{i=1}^{k} A^{-1}(d_i + X). \tag{8}$$

On one hand such an attractor is given explicity by the summation formula (3), and on the other hand it is, by comparing equations (5) and (8), the self-replicating tile for (L, A, D).

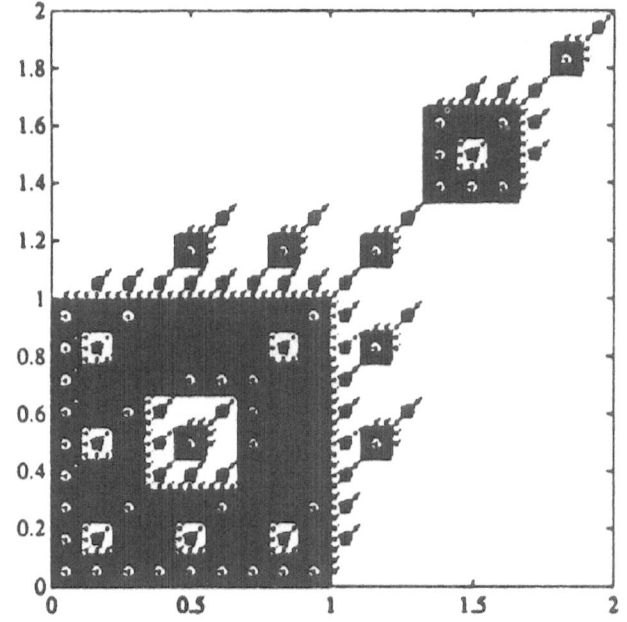

Figure 8. $T\left(\mathbb{Z}^2, \begin{pmatrix} 3 & 0 \\ 0 & 3 \end{pmatrix}, \left\{ \begin{pmatrix} 0 \\ 0 \end{pmatrix}, \begin{pmatrix} 1 \\ 0 \end{pmatrix}, \begin{pmatrix} 2 \\ 0 \end{pmatrix}, \begin{pmatrix} 0 \\ 1 \end{pmatrix}, \begin{pmatrix} 0 \\ 2 \end{pmatrix}, \begin{pmatrix} 1 \\ 2 \end{pmatrix}, \begin{pmatrix} 2 \\ 1 \end{pmatrix}, \begin{pmatrix} 2 \\ 2 \end{pmatrix}, \begin{pmatrix} 4 \\ 4 \end{pmatrix} \right\} \right).$

Figure 9. Approximations to the rep-2 dragon tile.

Theorem 1 naturally suggests the following two questions, which will be addressed in Section 5.

(1) Is it possible to determine whether a given radix system (L, A, D) has the unique factorization property?

(2) Given (L, A), does there exist at least one digit set D for which (L, A, D) has the unique representation property?

4. Generalized balanced ternary

Just as the decimal system is suitable for denoting integer points on the line, any radix system (L, A, D) with the unique representation property provides a system for addressing lattice points in Euclidean n-space, the *address* of a lattice point x given by the string of digits in the radix representation of x. The generalized balanced ternary, defined below, is a useful n-dimensional radix system that simultaneously generalizes the 1-dimensional balanced ternary and the 2-dimensional hexagonal system.

Let A_n^* denote the dual of the classical n-dimensional root lattice A_n. For our purposes A_n^* is the lattice in $\mathbb{R}^n$ generated by a set $\{v_0, \dots, v_n\}$ of vertices of a regular n-simplex with barycenter at the origin. So A_1^* is the integer lattice on the line, and A_2^* is the hexagonal lattice in the plane. The Voronoi region of A_1^*, A_2^*, and A_3^* is an interval, a hexagon, and a truncated octahedron, respectively. In general, the Voronoi region of A_n^* is a permutohedron, the n-dimensional polytope whose vertices (embedded in $\mathbb{R}^{n+1}$) consist of the $(n+1)!$ points obtained by permuting the coordinates of $(-n/2, (-n+2)/2, (-n+4)/2, \dots, (n-2)/2, n/2)$.

The lattice A_n^* is isomorphic, as an Abelian group, to the quotient $L = \mathbb{Z}[x]/(f)$, where $f(x) = 1 + x + \cdots + x^n$. Let $\omega = \bar{x}$, where the bar denotes the coset in L containing x. Since $\omega^{n+1} = 1$, the ring L acts somewhat like adjoining an $(n+1)$st root of unity to $\mathbb{Z}$. The isomorphism between A_n^* and L is obtained by mapping generators $v_i \mapsto \omega^i$, and extending linearly. In fact, because L is a ring, the lattice A_n^* also inherits a ring structure. Addition in A_n^* is usual vector addition. By abuse of language no distinction will be made between A_n^* and L.

Let $\beta = \bar{2} - \omega \in L$ be the base for radix representation in A_n^*. The matrix representing β, with respect to the generators $\{v_0, \dots, v_n\}$, is

$$
A_\beta = \begin{pmatrix}
2 & 0 & \cdots & 0 & -1 \\
-1 & 2 & \cdots & 0 & 0 \\
0 & -1 & \cdots & 0 & 0 \\
\vdots & \vdots & \ddots & \vdots & \vdots \\
0 & 0 & \cdots & -1 & 2
\end{pmatrix}.
$$

Let the digit set be $D = \{\varepsilon_0 + \varepsilon_1\omega + \cdots + \varepsilon_n\omega^n : \varepsilon_i \in \{0, 1\},$ not all $\varepsilon_i = 1\}$. Then $(\mathbf{A}_n^*, \mathbf{A}_\beta, D)$ has the unique representation property [18], and is called the *generalized balanced ternary* (GBT). The 1- and 2-dimensional cases are exactly the balanced ternary and hexagonal systems. In computer vision the pixel locations in an image can be thought of as the lattice points at the centers of Voronoi cells that tile the plane. A geometric advantage of pixel locations on a hexagonal grid, rather than the usual square grid, is that the hexagons are a reasonably accurate approximation to a circle. A computer software advantage of GBT addressing is that high throughput rates are achieved by performing addition and multiplication, as well as conversion of address to planar locations and vice-versa, in terms of the bit strings $\varepsilon_0\varepsilon_1 \cdots \varepsilon_n$ that represent the digits [18][29]. One firm [6] has developed a planar database management system based on the 2-dimensional GBT. Figure 10 shows all planar locations with addresses of at most three GBT digits and also the product $25.255 = 604$. For convenience, the addresses in this figure are given using base 7 digits instead of binary string digits. This is possible because there is a ring isomorphism

$$\Theta : \mathbf{A}_n^* \to \mathbb{Z}$$

that, in dimension 2, takes a radix β representation in $\mathbf{A}_2^*$ with digits D to a radix 7 representation in $\mathbb{Z}$ with digits $\{0, 1, 2, 3, 4, 5, 6\}$. In general, the isomorphism is given as follows. Let $q = 2^{n+1} - 1$; then

$$\Theta : \sum_{i=0}^m d_i\beta^i \to \sum_{i=0}^m \theta(d_i)q^i,$$

where $\theta : D \to \{0, 1, \ldots, q - 1\}$ is given by

$$\theta : \sum_{i=0}^n \varepsilon_i\omega^i \mapsto \sum_{i=0}^n \varepsilon_i 2^i.$$

The proof that this is an isomorphism follows from the facts that $\omega^{n+1} = 1$ and $\omega \equiv 2 \mod (\beta L)$.

5. Unique representation

This section contains three remarks concerning when a radix system (L, A, D) has the unique representation property. Since D is a digit set, uniqueness follows

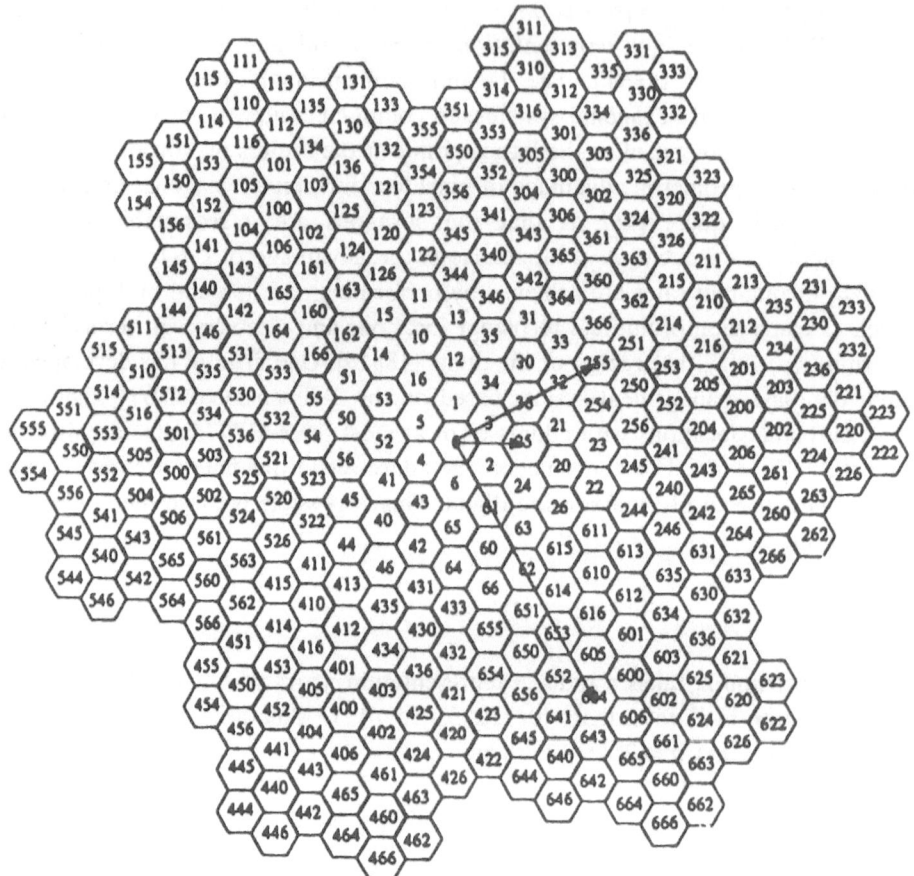

Figure 10. Addresses in generalized balanced ternary radix system.

from the proposition in Section 3. The issue is whether every lattice point has some representation. Proofs of the results in this section appear in [34].

REMARK 1. Assume in this remark that $A = bQ$ is a similarity; for the general case see [34]. Given a radix system (L, A, D), there is an efficient algorithm to determine whether (L, A, D) represents each lattice point. This algorithm is based on the following simple routine that is analogous to finding the base β digits of an integer.

Given a point $x = x_0 \in L$, define a sequence $\{x_i\}$ of lattice points and a sequence $\{d_i\}$ of digits recursively by the formulas

$$x_{i+1} = A^{-1}(x_i - d_i),$$

where d_i is the unique element of D such that

$$d_i \equiv x_i \quad \text{mod } A(L).$$

It is clear that, if $x_m = 0$ for some m, then $x_i = d_i = 0$ for all $i \geq m$. And, if there is a repetition in the sequence $\{x_i\}$ other than 0, then $x_i \neq 0$ for all i. Let $r(D) = \max\{|d| : d \in D\}/(b-1)$; let B be a ball centered at the origin with radius $|x| + r(D)$; and let α be the number of lattice points of L in B. Then it can be shown that x has a representation in (L, A, D) if and only if $x_m = 0$ for some $m \leq \alpha$. In this case, the digits in the radix representation of x are $d_0, d_1, \ldots, d_{m-1}$. Consider, for example, $(\mathbb{Z}, 3, \{-1, 0, 4\})$. If $x = 2$, then

$$x_0 = 2 \qquad d_0 = -1$$

$$x_1 = 4 \qquad d_1 = 4$$

$$x_2 = -1 \qquad d_2 = -1$$

$$x_3 = 0 \qquad d_3 = 0,$$

and 2 has base 3 representation $(-1)(4)(-1)$. If $x = -2$, then $x_0 = -2$ and $x_1 = -2$ so that -2 has no finite representation. Call this procedure for determining whether a given lattice point has a (finite) radix representation, *Algorithm A*.

ALGORITHM. *Every lattice point in L has a representation in the radix system (L, A, D) if and only if every lattice point in a ball of radius $r(D)$ has a finite radix representation. The latter condition can be efficiently tested using Algorithm A.*

As a first example, apply this algorithm to Example 3, the 2-dimensional GBT. In this case, $r(D) = 1/(\sqrt{7}-1)$. Since 0 is the only lattice point in the ball of radius $r(D)$, it is immediate that (L, A, D) has the unique representation property. Not quite so trivial, consider $L = \mathbb{Z}[i]$, $\beta = 1 + i$ and $D = \{0, 1\}$. In this case $b = \sqrt{2}$, so that $r(D) = 1/(\sqrt{2}-1)$. There are exactly 21 Gaussian integers in the ball of radius $r(D)$ to which Algorithm A is applied. However, Algorithm A applied to the Gaussian integer i (which lies in the ball) results in the sequence $i, i \ldots$. The repetition indicates that i has no finite radix representation. On the other hand with $\beta = -1 + i$, instead of $1 + i$, it can be checked using Algorithm A that all 21 lattice points in the ball do have finite radix representations. Therefore $(\mathbb{Z}[i], -1 + i, \{0, 1\})$ does have the unique representation property.

REMARK 2. A sufficient condition for representation of each lattice point can be stated in terms of the radix arithmetic.

SUFFICIENT CONDITION 1. *Every lattice point in L has a representation in the radix system (L, A, D) if*
(1) *each element in a basis for L has a radix representation, and*
(2) $D + D \subseteq D + A(D)$.

The second condition states that, in the radix arithmetic, a carry for addition can go at most one place to the left. Example 2 in Section 3, where digits are carried three places to the left, shows that this may not be the case. For the generalized balanced ternary, however, it can be shown that this is the case in all dimension.

REMARK 3. Given a lattice L and linear expansive map $A: L \to L$, does there exists *at least one set* D of digits such that (L, A, D) has the unique representation property? The answer, in general, is no. The simplest counterexample is the binary system for the integers: $L = \mathbb{Z}$ and $A = (2)$. In fact, this is the only example in dimension 1. In dimension 2 there are infinitely many counterexamples. Let $L = \mathbb{Z}^2$ and

$$A = \begin{pmatrix} 0 & -m \\ 1 & m \end{pmatrix} \quad \text{or} \quad A = \begin{pmatrix} 2 & m \\ 0 & 2 \end{pmatrix},$$

where m is an integer. In either case there is no set of digits so that $(\mathbb{Z}^2, A, D)$ has the unique representation property. In fact it can be proved that, if $\det(I - A) = \pm 1$, then there exists no appropriate set D of digits.

In the positive direction it can be shown that, if A has sufficiently large singular values, then there exists a digit set D such that (L, A, D) has the unique representation property. Recall that the singular value decomposition of a real matrix A is $A = U \circ \text{diag}(\sigma_1, \ldots, \sigma_n) \circ V^T$, where U and V are orthogonal matrices. The real numbers $\sigma_1, \ldots, \sigma_n$ are called the *singular values* of A. If $A = bQ$ is a similarity, then all singular values of A are equal to the expansion factor b.

SUFFICIENT CONDITION 2. *If all the singular values of A are greater than $3\sqrt{n}$, then there exists a set of digits D such that (L, A, D) has the unique representation property. In dimensions 1 and 2 the bound $3\sqrt{n}$ can be improved to 2.*

The digit set D in this result is set of lattice points contained in the image under A of the half open unit cube (a fundamental domain for the cubic lattice) centered at the origin. As an example consider the square lattice $L = \mathbb{Z}^2$ and the linear map

$$A = \begin{pmatrix} -2 & -1 \\ 1 & -2 \end{pmatrix}.$$

Both singular values of A are $2.2361 > 2$. Hence those lattice points $\{(0, 0), (1, 0), (0, 1), (-1, 0), (0, -1)\}$ that lie in the image under A of the half open square $(-\frac{1}{2}, \frac{1}{2}] \times (-\frac{1}{2}, \frac{1}{2}]$ constitute a digit set D for which (L, A, D) is the one that appears in Figure 5.

A generalization of Sufficient Condition 2 produces a family of digit sets D for each pair (L, A), each digit set being the image under A of the fundamental domain of a certain lattice [34]. Strichartz [31] proves a result similar to Sufficient Condition 2 for the case that A is a similarity. Gröchenig and Haas [14] construct a digit set D for each pair (L, A) in dimension 2 (except when A has two irrational real eigenvalues) that guarantees that $\mathcal{T}(L, A, D)$ is a lattice rep-tiling, And in some cases, they show that the rep-tile $T(A, D)$ is connected.

6. Variations and generalizations

This section discusses two open problems related to topics in this paper, and a generalization from lattice rep-tiles to crystallographic rep-tiles. The subject of self-replicating tilings using more than one prototile is fascinating, but too vast to be discussed here.

Two problems

In some of the questions below, the tile, rather than the tiling, is the fundamental object of study. A *rep-tile* is a compact set T in $\mathbb{R}^n$ of positive Lebesque measure with the property that there exists an expansive matrix A such that the image $A(T)$ is tiled by translates of T. As before, there is a set D consisting of $|\det A|$ digits defined by the equation $A(T) = \bigcup_{d \in D}(d + T)$, and $T = T(A, D)$ is given by formula (3) in Section 3. Note, however, that no lattice is mentioned in the definition of rep-tile. No tiling is mentioned either, but the following theorem holds [20]. Here $D_\infty = \bigcup_{i=1}^\infty D_m$, where $D_m = \sum_{i=0}^{m-1} A^i(D)$, and $\Delta(D_\infty) = D_\infty - D_\infty$.

THEOREM 2. *If $T(A, D)$ is a rep-tile, then there exists a set of translations $\mathcal{L} \subseteq \Delta(D_2)$ such that $\mathcal{L} + T(A, D)$ tiles $\mathbb{R}^n$.*

QUESTION 1. Is $T(L, A, D)$ a tiling even when the radix system (L, A, D) does not represent each lattice point?

Theorem 1 does not answer the question, because it does not even assert that L is a lattice, let alone the lattice L. It also does not assert that the tiling is a rep-tiling. (However, if $\mathscr{L} = L$ then it is easy to show that $T(L, A, D)$ is indeed a rep-tiling.) In fact, there is a counterexample to Question 1, even in dimension 1. Let $L = \mathbb{Z}$, $A = (3)$, $D = \{-2, 0, 2\}$. Then $T(A, D) = [-1, 1]$, so that adjacent tiles overlap on a unit interval.

Theorem 1 may, however, help in understanding this counterexample. If the lattice L in a radix system (L, A, D) has an A-invariant proper sublattice L' containing D, then clearly $\mathscr{L} \subseteq \Delta(D_\infty) \subseteq L' \subset L$. Since, by Theorem 1, $\mathscr{L} + T(A, D)$ tiles $\mathbb{R}^n$ it is impossible that $L + T(A, D)$ tiles $\mathbb{R}^n$. This is the case in the counterexample above, where the A-invariant sublattice containing D is $2\mathbb{Z}$. Of course, if L does have an A-invariant sublattice containing D, it is always possible to "mod out" and regard A as acting on the minimum (with respect to inclusion) such sublattice L'. So it is reasonable to conjecture, as was done in [14], that if L has no A-invariant sublattice containing D, then $T(A, D)$ does tile $\mathbb{R}^n$ by translation by L. The conjecture is true in dimension 1; however Lagarias and Wang [20] recently gave the following counterexample in dimension 2:

$$A = \begin{pmatrix} 2 & 1 \\ 0 & 2 \end{pmatrix} \qquad D = \left\{ \begin{pmatrix} 0 \\ 0 \end{pmatrix}, \begin{pmatrix} 3 \\ 0 \end{pmatrix}, \begin{pmatrix} 0 \\ 1 \end{pmatrix}, \begin{pmatrix} 3 \\ 1 \end{pmatrix} \right\},$$

where the tile $T(A, D)$ is shown in Figure 11. There is, however, a tiling (not a

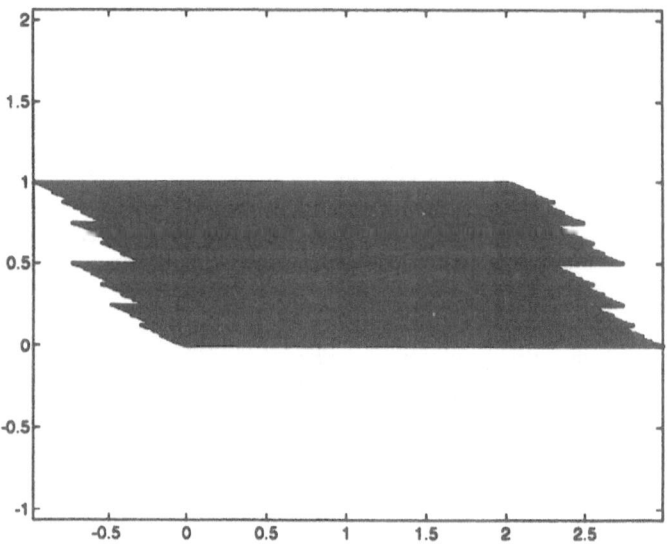

Figure 11. T does not give a tiling by translation by $\mathbb{Z}^n$.

rep-tiling) of $\mathbb{R}^2$ by translates of $T(A, D)$ to the lattice $3\mathbb{Z} \oplus \mathbb{Z}$. This motivates the weaker conjecture:

CONJECTURE 1. *If (L, A, D) is a radix system, then there is some lattice tiling using only translates of $T(A, D)$.*

Lagarias and Wang have announced a proof of Conjecture 1 in dimension 2 and in all dimensions when A is a similarity [21], but the general case appears to remain open. Concerning the original Question 1, Gröchenig and Haas [14] provide an algorithm that determines, given radix system (L, A, D), whether or not $\mathcal{T}(L, A, D)$ is a tiling.

Given any rep-tile T, Theorem 1 guarantees a tiling by translates of T but gives no limits on the degree of "disorder" the tiling can possess. This motivates the following question.

QUESTION 2. Does there exist a rep-tile that admits no periodic tiling? More generally, does there exist a region that tiles $\mathbb{R}^n$ by translation, but admits no periodic tiling?

In dimension 1, if a bounded region T tiles $\mathbb{R}$ by translation, then every tiling by translation is periodic [22]. The same result is true in dimension 2 if T is a topological disk with piecewise-C^2 boundary [11]. Likewise the result holds for convex polytopes in any dimension [33]. However, the Penrose tiles [27] are a set of two prototiles that admit infinitely many tilings of $\mathbb{R}^2$, but no periodic tilings. For one tile the answer to both questions, in general, is open. If translations are not the only motions allowed, then there are examples of both non-convex (Schmitt, unpublished) and convex (J. H. Conway, unpublished) polyhedra that tile $\mathbb{R}^3$ by Euclidean motions but only aperiodically.

Crystallographic rep-tiles

The group $\mathbb{Z}^n$, viewed as a discrete group of isometries acting on $\mathbb{R}^n$ with compact quotient, is just one of many such *crystallographic groups*. There are 7 in dimension 1 (frieze groups), 17 in dimension 2 (wallpaper groups); 230 in dimension 3; 4783 in dimension 4; and, in general, there are finitely many crystallographic groups in each dimension (proved by Bieberbach in 1910). The lattice rep-tilings are rep-tilings by all images of a single tile T under the action of the group $\mathbb{Z}^n$. Gelbrich [5] generalizes by calling a rep-tiling $\mathcal{T}$ crystallographic if $\mathcal{T} = \{\gamma(T) \mid \gamma \in \Gamma\}$, for some tile T and some crystallographic group Γ. He proves in dimension 2 that, for a crystallographic group and a given expansion factor k, there are finitely many isomorphism types of crystallographic rep-tiles that are homeomorphic to a disk. Here an isomorphism is an affine bijection that preserves tiles at every level in the hierarchy. Using an algorithm for determining the tiles, he shows, for example, that, in the case of a lattice rep-tile (crystallographic group generated by two independent translations), there are three classes of tiles for $k = 2$ and seven classes for $k = 3$.

REFERENCES

[1] BANDT, C., *Self-similar sets 5. Integer matrices and fractal tilings of* $\mathbb{R}^n$. Proc. Amer. Math. Soc. *112* (1991), 549–562.
[2] BARNSLEY, M., *Fractals everywhere*. Academic Press, Boston, 1988.
[3] DEKKING, F. M., *Recurrent sets*. Adv. Math. *44* (1982), 78–104.
[4] DEKKING, F. M., *Replicating superfigures and endomorphisms of free groups*. J. Combin. Theory Ser. A *32* (1982), 315–320.
[5] GELBRICH, G., *Crystalloagraphic reptiles*. Preprint.
[6] GIBSON L. and LUCAS, D., *Spatial data processing using generalized balanced ternary*. In *Proceedings of the IEEE Computer Society Conference on Pattern Recognition and Image Processing*. IEEE Computer Society, New York, 1981, pp. 566–571.
[7] GILBERT, W. J., *Fractal geometry derived from complex bases*. Math. Intelligencer *4* (1982), 78–86.
[8] GILBERT, W. J., *Geometry of radix representations*. In *The geometric vein: the Coxeter festscrift*. Springer, New York–Berlin, 1981, pp. 129–139.
[9] GILBERT, W. J., *Radix representations of quadratic fields*. J. Math. Anal. Appl. *83* (1981), 264–274.
[10] GILES, J., *Construction of replicating superfigures*. J. Combin. Theory Ser. A *26* (1979), 328–334.
[11] GIRAULT-BAUQUIER F. and NIVAT, M., *Tiling the plane with one tile*. In *Topology and category theory in computer science* (G. M. Reed, A. W. Roscoe and R. F. Wachter, eds.), Oxford Univ. Press, 1989, pp. 291–333.
[12] GOLOMB, S. W., *Replicating figures in the plane*. Math. Gaz. *48* (1964), 403–412.
[13] GRÖCHENIG K. and MADYCH, W. R., *Multiresolution analysis, Haar bases and self-similar tilings of* $\mathbb{R}^n$. IEEE Trans. Inform. Theory *38* (1992), 556–568.
[14] GRÖCHENIG K. and HAAS, A., *Self-similar lattice tilings*. Preprint.
[15] GRÜNBAUM, B. and SHEPHARD, G. C., *Tilings and patterns*. W. H. Freeman and Company, New York, 1987.
[16] KÁTAI, I. and SZABÓ, I., *Canonical number systems for complex integers*. Acta Sci. Math. (Szeged) *37* (1975), 255–260.
[17] KENYON, R., *Self-replicating tilings*. In *Symbolic dynamics and its applications* (P. Walters, ed.). [Contemp. Math., Vol. 135]. Birkhäuser, Boston, 1992, pp. 239–264.
[18] KITTO, W., VINCE A. and WILSON, D., *An isomorphism between the p-adic integers and a ring associated with a tiling of n-space by permutohedra*. Discrete Appl. Math. *52* (1994), 39–51.
[19] KNUTH, D. E., *The art of computer programming, Vol. 2, Seminumerical algorithms*. 2nd ed. Addison-Wesley, Reading, Mass., 1981.
[20] LAGARIAS, J. C. and WANG, Y., *Self-affine tiles in* $\mathbb{R}^n$. Preprint.
[21] LAGARIAS, J. C. and WANG, Y., *Integral self-affine tiles in* $\mathbb{R}^n$: *Standard and nonstandard digit sets, II. Lattice tiling*. Preprint.
[22] LAGARIAS, J. C. and WANG, Y., *Tiling the line with one tile*. Preprint.
[23] LAWTON, W. and REESNIKOFF, H. L., *Multidimensional wavelet bases*. Preprint.
[24] MANDELBROT, B. B., *The fractal geometry of nature*. Freeman, San Francisco, 1982.

[25] MATULA, D. W., *Basic digit sets for radix representations.* J. Assoc. Comput. Mach. *4* (1982), 1131–1143.

[26] ODLYZKO, A. M., *Non-negative digit sets in positional number systems.* Proc. London Math. Soc. *37* (1978), 213–229.

[27] PENROSE, R., *Pentaplexity.* Math. Intelligencer *2* (1979), 32–37.

[28] RADIN, C., *Symmetry of tilings of the plane.* Bull. Am. Math. Soc. *29* (1993), 213–217.

[29] VAN ROESSEL, J. W., *Conversion of Cartesian coordinates from and to generalized balanced ternary addresses.* Photogrammetric Eng. Remote Sensing *54* (1988), 1565–1570.

[30] SHECHTMAN, D., BLECH, I., GRATIAS, D. and CAHN, J., *Metallic phase with long-range orientational order and no translational symmetry.* Phys. Rev. Lett. *53* (1984), 1951–1954.

[31] STRICHARTZ, R. S., *Wavelets and self-affine tilings.* Constructive Approx. *9* (1993), 327–346.

[32] THURSTON, W., *Groups, tilings and finite state automata.* [AMS Colloquium Lecture Notes]. Amer. Math. Soc., Providence, RI, 1989.

[33] VENKOV, B. A., *On a class of Euclidean polyhedra.* Vestnik Leningrad. Univ. Mat. Fiz. Khim. *9* (1954), 11–31 (Russian).

[34] VINCE, A., *Replicating tessellations.* SIAM J. Discrete Math. *6* (1993), 501–521.

Department of Mathematics,
University of Florida,
Gainesville, Florida 32611,
U.S.A.

BIRKHÄUSER

A. Joseph / F. Mignot / F. Murat / B. Prum / R. Rentschler (Eds)

First European Congress of Mathematics

Paris, July 6–10, 1992

The three volume work containing the proceedings of the first European Congress of Mathematics encompasses an account of the state of research in a wide variety of mathematical topics, as well as broad ranging discussions of the role of mathematics in society.
Volumes I and II form a collection of the manuscripts contributed by the invited lecturers. Volume III contains the Round Table report.

Vol. I: Invited Lectures (Part 1)
(PM 119)
1994. 594 pages. Hardcover

Contributors:
V. I. Arnold, Z. Adamowicz,
L. Babai, A. Björner,
C. De Concini, B. Bojanov,
S. K. Donaldson, J.-M. Bony,
W. Müller, R. E. Borcherds,
D. Mumford, J. Bourgain,
A.-S. Sznitman, F. Catanese,
M. Vergne, C. Deninger,
S. Dostoglou and D. Salamon

Vol. II: Invited Lectures (Part 2)
(PM 120)
1994. 545 pages. Hardcover

Contributors:
D. Duffie, M. A. Nowak,
J. Fröhlich, R. Piene,
M. Giaquinta, A. Quarteroni,
U. Hamenstädt, A. Schrijver,
M. Kontsevich, B. Silverman,
S. B. Kuksin, V. Strassen,
M. Laczkovich, P. Tukia,
J.-F. Le Gall, C. Viterbo,
I. Madsen, D. Voiculescu,
A. S. Merkurjev, M. Wodzicki,
J. Nekovár, D. Zagier, Y. Neretin

Vol. III: Round Tables (PM 121)
1994. 608 pages. Hardcover

Topics:
Mathematics and the general public • Women and mathematics • Mathematics and educational policy • Let's cultivate mathematics! • Mathematical Europe: Myth or historical reality? • Philosophie des mathématiques : pourquoi ? comment ? • Mathématiques et sciences sociales • Mathematics and industry • Degree harmonization and student exchange programmes • The Pythagoras programme • Collaboration with developing countries • Mathematical libraries in Europe • Mathematics and economics • Mathématiques et chimie • Mathematics in medicine and biology

Set Vols 1–3
ISBN 3-7643-2801-0

Volume III (Round Tables) also available as softcover edition.

1994. 608 pages. Softcover
ISBN 3-7643-5156-X

Please order through your bookseller or write to:
Birkhäuser Verlag AG
P.O. Box 133
CH-4010 Basel / Switzerland
FAX: ++41 / 61 / 271 76 66
e-mail: 100010.2310@compuserve.com

For orders originating in the USA or Canada:
Birkhäuser
333 Meadowlands Parkway
Secaucus, NJ 07094-2491 / USA

Birkhäuser

Birkhäuser Verlag AG
Basel · Boston · Berlin

W. Ballmann, University of Bonn, Germany

Lectures on Spaces of Nonpositive Curvature
with an appendix by Misha Brin
Ergodicity of Geodesic Flows

DMV Seminar 25 1995. Approx. 120 pages. Softcover
ISBN 7643-5242-6

Singular spaces with upper curvature bounds and, in particular, spaces of nonpositive curvature, have been of interest in many fields, including geometric (and combinatorial) group theory, topology, dynamical systems and probability theory. In the first two chapters of the book, a concise introduction into these spaces is given, culminating in the Hadamard-Cartan theorem and the discussion of the ideal boundary at infinity for simply connected complete spaces of nonpositive curvature.

In the third chapter, qualitative properties of the geodesic flow on geodesically complete spaces of nonpositive curvature are discussed, as are random walks on groups of isometries of nonpositvely curved spaces. The main class of spaces considered should be precisely complementary to symmetric spaces of higher rank and Euclidean buildings of dimension at least two (Rank Rigidity conjecture). In the smooth case, this is known and is the content of the Rank Rigidity theorem. An updated version of the proof of the latter theorem (in the smooth case) is presented in Chapter IV of the book. This chapter contains also a short introduction into the geometry of the unit tangent bundle of a Riemannian manifold and the basic facts about the geodesic flow.

Bitte bestellen Sie bei Ihrem   Für Bestellungen aus den USA
Buchhändler oder direkt bei: oder Canada: *Birkhäuser*

Birkhäuser Verlag AG   Birkhäuser
P.O. Box 133   333 Meadowlands Parkway   Birkhäuser Verlag AG
CH-4010 Basel / Schweiz   Secaucus, NJ 07094-2491   Basel · Boston · Berlin
FAX: ++41 / 61 / 271 76 66   USA